Wargamers Handbook of the American War of Independence

by

DONALD FEATHERSTONE

Member of The International Society of Military Collectors
The Society for Army Historical Research
The Military Historical Society
The Society of Friends of the National Army Museum
The Wessex Military Society
The British Model Soldier Society
The Society of Ancients
The Pike and Shot Society
The Model Generals Club of America
The International Federation of Wargaming of America
The British Commission of Military Historians

Donald Featherstone

Wargamers Handbook of the American War of Independence

Model and Allied Publications
Argus Books Limited
14 St James Road, Watford
Hertfordshire, England

First published 1977

ISBN 0 85242 544 9

Printed in Great Britain
by W & J Mackay Limited, Chatham

CONTENTS

ACKNOWLEDGEMENTS

Acknowledgements and thanks are given to the well-known firm of model soldier manufacturers TRADITION of London, on whose premises all the photography of the individual models was carried out, using figures manufactured by them and painted by their artists.

The dust-jacket illustrations, showing uniforms of the period, are reproduced by courtesy of *Military Modelling* magazine.

PREFACE

THE AMERICAN War of Independence, some two hundred years ago, was one of the great cataclysmic events of history. It took the British colonies in America out of British hands, and saw the birth of the United States of America. In military terms the war and its battles has always had much of interest to offer the model soldier and wargames enthusiast. There was a great diversity of units and uniforms, new styles of fighting, and many battles on a pleasingly small scale without too much complexity. The period lends itself admirably to wargaming on table-top battlefields, and authentic figures and equipment are readily available through the hobby trade. This book is a complete manual for wargamers on the events of the period, the battles, and the men. It includes concise but comprehensive rules, the pick of the most suitable battles to refight in miniature, and a discussion on the wargame aspects of each battle. All the essential facts and figures are presented in easy to follow form, and the final sections give much practical modelling information and a bibliography of useful books for further reading. Though written in the first instance specifically for wargamers this book should be of value equally to non-wargamers who will find the lucid information on the war, the battles, the commanders, and the men of great value in any military study of this fascinating period.

THE WAR OF THE AMERICAN REVOLUTION— 1775/1783

LONG STANDING DISAGREEMENTS between Britain and the thirteen Colonies flamed into open rebellion at Lexington in April 1775 when 700 soldiers from the British garrison at Boston opened fire on a gathering of Colonists. Then, after destroying supplies at Concord, the British had their first taste of Colonial guerrilla warfare as they marched back to Boston harassed by swarms of Militiamen, hidden behind trees and stonewalls who poured musket fire into their packed ranks. Two months later on Bunker Hill overlooking Boston Harbour 2,200 British Regular troops lost practically half their number when repeatedly assaulting earthworks manned by American militiamen. Throughout the war the Americans made greater use of entrenchments than most armies of the period. Tactically a British victory, Bunker Hill gave the Americans tremendous pyschological encouragement, turning their superstitious awe of the Regulars into a characteristic indisciplined and brash over confidence.

In July 1775 George Washington became commander of the Continental forces and began his monumental task of transforming an armed rabble into an army. Among other handicaps, he was plagued by a shortage of trained officers, the best American leaders were in the Northern campaigns. The remainder of that year and the first half of 1776 saw an inadvisable and ill-equipped attempt to invade Canada by small forces under Benedict Arnold and General Montgomery who, after capturing Montreal, were repulsed at Quebec and eventually withdrew from Canadian soil.

In July 1776 General Howe sailed from Halifax and landed in New York harbour with 32,000 men including 9,000 Hessian infantry.*

*During the American Revolution, the British hired about 32,000 mercenaries from Hesse-Cassell; Hesse-Hanau; Brunswick/Luneburg; Waldeck; Ansbach-Bayreuth; Anhalt-Zerbst and Hanover. These troops were not

Out-manoeuvred in the Battle of Long Island, Washington saved his army by a brilliant withdrawal when Glover's Amphibious Regiment of Massachusetts fishermen ferried every single man to safety. Later, they acquired a reputation fighting as gunners or infantrymen, when the army did not require their special skill in inland water operations.

With his little army growing smaller as enlistments expired, Washington retreated through New Jersey, pausing to lead a force of 2,400 men across the River Delaware on Christmas night 1776 and completely destroy the Hessian garrison at Trenton. Most armies went into winter quarters during hard weather so that any general or army that was willing or able to fight during the winter gained surprise successes. Lord Cornwallis rushed up 5,000 Regulars to trap Washington's 1,600 Regulars and 3,600 Militia, who escaped by leaving camp fires burning and slipping away in the night. Washington next defeated British reinforcements advancing from Princetown, to conclude a brilliant ten day campaign that 'fanned the dying embers of the Revolution into a lively flame'.

In early 1777, General Burgoyne, General Howe and Colonel St. Leger, aided by Brant's Indians and Johnson's Tories, embarked on a three-pronged offensive which demanded close co-ordination and timing, difficult to achieve over the rough trails and in the trackless forests of North America. Burgoyne caputured Fort Ticonderoga in July, as Howe in New Jersey moved out against Washington; who refused battle while Morgan's riflemen harassed Howe's flanks. Frontiersmen, such as those led by Morgan, operated independently in groups. Their long rifle, proven in many Indian wars, was a vastly superior weapon to the smooth-bore musket which could only fire with 40% accuracy at 100 yards against the rifle's 50% accuracy at 300 yards. By using a greased patch around the ball range, accuracy and penetration were improved. However it took twice as long to load and, unable to carry a bayonet, was useless against a bayonet attack and could not stand ground when rapid fire was required.

By drawing Howe southwards, Washington did much to prevent his linking up with Burgoyne, although he was defeated at Brandywine and Germantown, where it was shown that short-term American volunteers were not yet ready to meet the Regulars in formal combat, falling easy victims to bayonets that were on them before they could re-load. The musket had a 20% hit probability at 150 yards, but infantry could charge 100 yards in a minute so that troops close enough to suffer casualties from musketry were also within charge-range. The psychological effect of the bayonet brought the

---

particularly good and desertion was common; at no time were there more than 20,000 of them in the Western hemisphere.

British many victories during the American Revolution.

Burgoyne's force consisted of 7,200 troops including 3,000 German mercenaries and 138 guns in the biggest artillery train seen in America plus an Indian and Canadian scouting arm that tactically blinded Gate's army. The American militiamen, knowing nothing of frontier fighting methods, were terrified of the Indians but the arrival of Morgan's riflemen in early September reinforced Gate's outnumbered, demoralised and badly led army and they turned the tables by outwitting Burgoyne's scouts, causing his army to grope.

St. Leger on the Mohawk with 875 British, Tory and Hessian troops and 1,000 Iroquois Indians under Brant, besieged Fort Schuyler and at Oriskany ambushed Herkimer's relieving force but later the fort was relieved by Benedict Arnold, when St. Leger retreated. On 16th August, a force of 2,000 militia plus 400 Green Mountain Boys wiped out two forces of Brunswickers at Bennington, killing 207 and taking 700 prisoners out of a total of 1,350 men.

By September, Burgoyne knew that Howe was not going to join him as planned so decided to strike out for Albany, hoping to meet St. Leger. Crossing the Hudson, Burgoyne attacked the Americans at Freeman's Farm—a tactical landmark because riflemen bore the brunt of the battle. Morgan's men in brown linen hunting shirts and buckskin breeches, almost invisible amid the Autumn colouring of the woods, took heavy toll of the scarlet-clad British infantry advancing in close order across forest clearings. The battle swayed backwards and forwards in the forest before ending at dusk in a tactical draw.

Burgoyne entrenched his army on Bemis Heights, hoping for help from Clinton's force in New York, but this commander did little except create a minor diversion up the Hudson before retiring back to New York. So, on 7th October, Burgoyne made a desperate attempt to break out, only to be driven back, with 600 casualties, by a counter-attack led by Arnold with Morgan's riflemen playing a conspicuous part. Withdrawing to Saratoga, Burgoyne's 5,700 men were surrounded by a force three times their strength, and constantly harassed by riflemen. On 17th October 1777 came the turning point of the Revolution when Burgoyne surrendered on the promise of his army being paroled back to Britain, but this was not honoured.

Under conditions of great hardship Washington and his army spent the winter of 1777/78 at Valley Forge where Baron von Steuben painstakingly taught the starved and cold little Continental army discipline and tactical skill. At the same time, Legions of Light Horse were raised by Harry Lee and William Washington, to aid the militiamen who up to now had been at the mercy of the British Dragoons. They were organised as four units of 50 men each, usually combining as a corps.

In June 1778, Sir Henry Clinton took over command from Howe and, evacuating Philadelphia, led his army of 10,000 redcoats (the baggage train was ten miles long) back to New York, followed by Washington's army on a parallel route. Clinton turned to give battle at Monmouth, which ended in an empty American victory although the Continental troops proved themselves equal to the British Regulars and, in a fair fight with musket and bayonet, Wayne's men shatttered a Guard Regiment.

Throughout 1779 and the following year, Washington blockaded the British in New York during a dreary period when, with his army slowly disintegrating from expired enlistments, only the personality of their leader kept the American cause alight. There were no major operations but some hard-fought minor actions and guerrilla warfare such as Wayne's recapture of Stony Point in July 1779.

For nearly two years there had been continuous guerrilla warfare between Tories and patriots in the Southern colonies but no large-scale fighting until Clinton moved to conquer the Carolinas. In December 1778, the British captured Savannah and its garrison under General Prevost successfully withstood siege in September/October 1779 when a large French force, with Continentals and militia, invested the town; the final French assault on October 8th was thrown back after hand-to-hand fighting in the most severe conflict of its type since Bunker Hill. The French Commander d'Estaing embarked his troops and sailed away, leaving Savannah in British hands. Leaving New York under the command of the German General von Knyphausen in December 1779 Clinton sailed with 8,000 men to attack Charleston, which was besieged from February until May 1780. The surrender of 5,400 men and great quantities of arms and equipment was considered the worst American disaster of the war. Clinton returned to New York leaving Cornwallis with 8,000 men to pacify South Carolina where guerrilla warfare involving Banastre Tarleton's Tory Cavalry Legion went on throughout the summer. In August 1780 General Gates' army of 3,000 militia and 900 Continentals were overwhelmed at Camden by Cornwallis with 2,400 British and Tory Regulars; Tarleton's Cavalry crushed the Continentals in a rear attack to begin the American rout. Considering that Militia were a disorganised, ill-trained and ill-disciplined group and were usually deployed in skirmisher lines or purely used for defensive purposes, it is not surprising that they were defeated by a smaller Regular force.

Washington sent General Greene with a force of his best Continentals to check Cornwallis and a victory was gained at Kings Mountain in October 1780 when British Colonel Patrick Ferguson's Corps of Tory Riflemen were shot to pieces by a slightly larger force of mountain men sharpshooters in a fight that was remarkable because both forces were composed entirely of Americans. By mid-December 1780 Green's army had grown to 3,000, including 1,400 Continentals while Cornwallis led a force of 4,000 well equipped and trained Regulars. In January 1781, Greene split his forces and sent General Morgan with 1,000 men off on a wide western sweep, chased by Tarleton who was decisively defeated at Cowpens in what was perhaps the most brilliant tactical operation ever fought on American soil. Morgan placed his unreliable militiamen in front, with permission to retire after they had fired two rounds; behind them were his Continentals with 80 Dragoons. Tarleton's Regulars, taking casualities from the two volleys fired by the Militia before they fled to the cover of a low hill in the rear, came to grips with the Continentals who made a desperate stand with musket and bayonet. Meanwhile Morgan had rallied his militia and led them entirely round the field, under cover of the ridge, to hit the British in their left rear. Simultaneously, the American Dragoons struck their right and eventually only Tarleton and a few scattered riders escaped. The Americans had about 70 killed and wounded; the British 110 killed and 830 captured out of their original 1,100.

Greene retreated into Southern Virginia closely pursued by Cornwallis who had destroyed his baggage in order to increase his mobility. In March 1781 at Guilford Courthouse Greene's army of 4,400 men, two-thirds inexperienced militia, turned to face Cornwallis' 1,900 Regulars. In wooded country, Greene positioned his troops in three lines; in turn each was driven back by the outnumbered British who, in spite of heavy losses, fought with great gallantry. Greene broke off the action and retreated after inflicting such casualities on Cornwallis that he had to abandon the campaign and withdraw into Virginia. Greene attacked several isolated British garrisons in the Carolinas at Hobkirk's Hill, Fort Ninety-six and Eutaw Springs, suffering a series of tactical defeats which nevertheless led to a strategic victory. Although his strength seldom exceeded 3,000 men, Greene had forced the invaders out of the South mainly because of the 1,600 Continentals who formed the backbone of his army. Unpaid, ill-equipped and always hungry, these men marched hundreds of miles and fought at least four major actions and innumerable lesser affairs in proving themselves the equal of the British Regular.

In July 1780, Rochambeau with 5,000 French troops landed at Newport but a British naval blockade prevented operations against New York, where Washington maintained his blockade despite four mutinies in his army during 1780/1. General Benedict Arnold defected to the British in September 1780.

In April 1781, Washington sent Lafayette with 3,500 men (including 1,200 veteran Continentals) and Wayne with 1,000 Continentals to Virginia where for several months the pair skilfully manoeuvred against the superior numbers of Cornwallis, who had been reinforced. In July at Jamestown Ford Wayne, caught by surprise, rallied and repulsed the British, although he took heavy losses. The French force under Rochambeau moved from Newport and joined Washington outside New York as Cornwallis marched with 7,000 troops to Yorktown in August, watched by Lafayette with 4,500 at nearby West Point, Virginia, where he was reinforced with 3,000 French troops brought by the fleet of Admiral de Grasse. French battalions were organised into nine companies—eight companies were fusiliers and one grenadier company, with a 'paper' strength of about 740 although French regiments were usually nearer to their establishment than British, Americans or Hessians. British Admiral Graves, badly handling his 19 ships against de Grasse's 24 at the Battle of the Capes, withdrew to New York so giving command of the sea to the French and sealing Cornwallis's fate.

Leaving a mere 2,000 men to contain Clinton's 17,000 in New York, Washington forced-marched his army and the French force under Rochambeau, embarking them at the head of the Chesapeake in French ships to reach Yorktown on September 28th. The Allied French and American army was formed of some 9,500 Americans and 7,800 French Regular soldiers with a good artillery train of field and siege-guns.

Hoping for aid from Clinton, Cornwallis withdrew his 8,000 men into the inner fortifications at Yorktown, so allowing the Allies to bring their siege artillery within range of his entire position. After a four day bombardment, the Allies stormed two important redoubts, established new batteries and repulsed a British counter-attack. His position impossible, Cornwallis surrendered on October 19th 1781, to virtually end the American Revolution. In November 1782, the Treaty of Paris was signed concluding the war and recognising the independence of the United States. The last British troops left New York in November 1783.

EASY REFERENCE CHARTS

—The War in the North
—The War in the South

IN THE NORTH—AMERICAN

Date	Battles	Commanders	Force	Casualties
1775				
April 19	Lexington & Concord	Parker, Barrett & Heath	3,763 in all	49k & 41w
June 17	Bunker Hill (Breed's Hill)	Gen. Putman, Cols. Prescott & Stark	about 2,000	140k, 271w, 30c.
Sept. 5 to Nov. 2	Siege of St. Johns	Gen. Richard Montgomery	1,200 to 1,500	
Oct. 17	Falmouth	Town bombarded and burned by 2 British warships		
Oct. 25	Montreal	Ethan Allen	110	about 40c
Dec. 31	Attack on Quebec	Gen. Montgomery, Col. Arnold	800	60k & w, 426c
1776				
May 15–20	The Cedars	Butterfield & Sherburne	540	400c
June 8	Trois Rivieres	Gen. William Thompson	2,000	160k & w, 236c
Aug. 27	Long Island	Gens. Washington, Putman, Sullivan & Sterling	about 3,500 out of 19,000	about 300k & 1,100w or c
Sept. 15	Kip's Bay	Col. William Douglas	about 900	60k & w, 320c
Sept. 16	Harlem Heights	Washington, Greene, Putman	about 2,000	30k & 100w
Oct. 11–13	Valcour Island	Gens. Benedict Arnold, & David Waterbury	750 men in 15 vessels	80k & w and 11 vessels
Oct. 18	Pell's Point	Col. John Glover	750	8k & 13w
Oct. 22	Mamaroneck	Col. John Haslet	750	3k & 12 w
Oct. 28	White Plains	Gens. Washington, McDougall	about 1,600 out of 14,500	28k & 126w
Nov. 16	Fort Washington	Col. Robert Magaw	2,900	53k, 96w, 2,218c
Dec. 26	Trenton	General Washington	2,400	4w
1777				
Jan. 3	Princetown	General Washington	about 4,000	30k & 75w
April 27	Ridgefield (Danbury Raid)	Arnold, Wooster & Silliman	700	20k & 80w
June 26	Metuchen	General Lord Stirling	2,200	12k, 50w & 50c
July 7	Hubbardton	Col. Seth Warner	730	41k, 95w & 234c
July 8	Fort Ann	Cols. Long & Van Rensselaer	550	several k & w, 30c
Aug. 2–23	Siege of Fort Stanwix	Cols. Gansevoort & Willet	750	
Aug. 6	Oriskany	Gen. Nicholas Herkimer	860	about 200k & w
Aug. 16	Bennington	Gen. Stark & Col Warner	2,330	30k & 50w
Aug. 22	Staten Island (Raid)	General John Sullivan	about 1,000	10k, 15w & 140c
Sept. 3	Cooch's Bridge	General William Maxwell	720	about 40k & w
Sept. 11	Brandywine	General Washington	11,000	200k, 750w, 400c
Sept. 19	Freeman's Farm (Saratoga)	Gens. Gates & Arnold & Col. Daniel Morgan	about 3,000 out of 7,000	65k & 218w
Sept. 21	Paoli	General Anthony Wayne	1,500	150k & w, 17c
Oct. 4	Germantown	General Washington	11,000	152k, 521w, 400c
Oct. 6	Forts Clinton & Montgomery	Gens. George & James Clinton	at least 600	25k &w, 227c

IN THE NORTH—BRITISH

Commanders	Force	Casualties	Result
Lord Percy & Lt. Col. Smith	1,800	73k & 174w	American victory
Gens. Howe, Pigot & Clinton	2,400	226k & 828w	British victory
Major Charles Preston	725	725c	American victory
under Capt. Henry Mowat, 417 buildings & 11 vessels destroyed; 4 vessels taken.			
Gen. Guy Carleton	235		British victory
Gen. Guy Carleton	1,800	5k & 13w	British victory
Captain Forster	640		British victory
Gens. Carleton & Fraser	6,000	8k & 9w	British victory
Gens. Howe, Clinton, Grant, Cornwallis & Heister	about 17,000 out of 32,000	63k, 314w & 23c	British victory
Howe, Clinton, Cornwallis	4,000 plus	about 12k & w	British victory
Gen. Alexander Leslie	about 5,000	14k & 157w	Tactical draw
Gen. Guy Carleton & Capt. Thomas Pringle	1,670 men in 20 vessels	about 40k & w and 3 vessels	British victory
Gens. Howe & Cornwallis	4,000	at least 25k & w	British victory
Major Robert Rogers	428	several k, w, 36c	American victory
Gens. Howe, Leslie & Heister	about 4,000 out of 14,000	313k & w	British victory
Gen. Knyphausen	8,000	78k & 374w	British victory
Col. Johann Rall	1,400	22k, 92w & 948c	American victory
Lt. Col. Charles Mawhood	1,200	60k, 150w & 244c	American victory
Tryon, Agnew & Erskine	2,000	154k, & w, 40c	British victory
Gen. Charles Cornwallis	4,000	70k & w	British victory
Gens. Fraser & Riedesel	1,030	60k & 148w	British victory
Lt. Col. John Hill	190	13k, 23w & 15c	Tactical draw
Lt. Col. St. Leger & Jos. Brant	1,875		American victory
John Butler & Joseph Brant	about 1,000	about 150k & w	British victory
Lt. Cols. Baum & Breymann	1,442	207k & 700c	American victory
Gen. John Campbell	about 3,000	259c	British victory
Lt. Col. Ludwig von Wurmb		at least 30k & w	British victory
Howe, Cornwallis, Knyphausen	12,500	90k & 480w	British victory
General John Burgoyne	about 3,000	600k, w & c	American victory
General Charles Grey		4k & 5w	British victory
Gen. Sir William Howe	9,000	70k, 450w & 14c	British victory
Gen. Sir Henry Clinton	2,100 out of 3,000	40k & 150w	British victory

IN THE NORTH—AMERICAN

Date	Battles	Commanders	Force	Casualties
Oct. 7	Bemis Heights (2nd Saratoga)	Gens. Gates & Arnold & Col. Daniel Morgan	about 9,200 out of 1,,000	50k & 150w
Oct. 22	Fort Mercer (Red Bank)	Greene & Hazelwood	about 400	14k & 23w
Nov. 10–15	Fort Mifflin (bombardment)	Col. Smith & Maj. Thayer	450	250k & w.
Dec. 5–8	White Marsh	Gens. Washington & Irvine	11,000	about 90k & w, 32c
1778				
June 28	Monmouth	General Washington	about 11,000 out of 13,425	142k, 300w, 37 died of sunstroke
July 3	Wyoming	Col. Zebulon Butler	360	about 300k or c
Aug. 29	Quaker Hill (Newport)	Gen. John Sullivan	1,500 of 5,000	30k & 137w
Nov. 11	Cherry Valley	Col. Ichabod Alden	at least 250	70k & 33c
1779				
July 16	Stony Point	Gen. Anthony Wayne	1,350	15k & 83w
July-Aug.	Penobscot Expedition	Gens. Lovell & Wadsworth & Capt. Saltonstall	1,000 men & 40 ships	474k, w & c all 40 ships
Aug. 19	Paulus Hook	Major Henry Lee	300	2k & 3w
Aug. 29	Newtown (Elmira)	Gen. John Sullivan	3,462	3k & 39w
1780				
Feb. 3	Young's House	Lt. Col Joseph Thompson	450	14k, 37w & 76c
June 7 & 23	Springfield	Greene, Maxwell, Dickinson	at least 1,800	about 15k, 61w
Oct. 19	Klock's Field	Gen. Robt. Van Rensselaer	about 1,500	
1781				
Sept. 6	Fort Griswold	Lt. Col.'William Ledyard	164	85k & 60w
Oct. 25	Johnstown	Col. Marinus Willett	400	40k & w

Casualties are indicated by the following initials: k = killed; w = wounded; c = captured; d = deserted.

IN THE SOUTH—AMERICAN

Date	Battles	Commanders	Force	Casualties
1775				
Dec. 9	Great Bridge	Col. William Woodford & Lt. Edward Travis	at least 90 out of 1,000	1w
1776				
Feb. 27	Moore's Creek Bridge	Cols. Caswell & Lillington	1,000	1k & 1w
June 28	Fort Sullivan	Col. William Moultrie	435	17k & 20w
1778				
Dec. 29	Savannah	General Robert Howe	850	83k & 453c
1779				
Feb. 3	Beaufort (Port Royal Island)	Gen. William Moultrie	320	8k & 22w
Feb. 14	Kettle Creek	Col. Andrew Oickens	300	9k & 23w

IN THE NORTH—BRITISH

Commanders	Force	Casualties	Result
Gens. Burgoyne, Fraser and Riedesel	about 2,200 out of 5,700	176k, 250w & 200c	American victory
Col. Carl von Donop	1,200	377k & w, 20c	American victory
Gen. Howe & Adm. Howe	6 ships; btyrs.	13k, 24w; 2 ships	British victory
Gens. Howe & Cornwallis	14,000	about 60k &w	American victory
Gen. Sir Henry Clinton	9,500 out of 13,000	190k, 390w, 576d 59 died of sunstroke	Tactical draw
Colonel John Butler	900	3k & 8w	British victory
General Sir Robert Pigot	3,000	38k & 21w	British victory
Walter Butler, Joseph Brant	700		British victory
Lt. Col. Henry Johnson	625	20k, 74w & 543c	American victory
Col. Francis MacLean	600	12k & w	British victory
Maj. William Sutherland	about 250	50k & w, 158c	American victory
J. & W. Butler & Jos. Brant	600 to 1,200	at least 12k	American victory
Lt. Col. Chapple Norton	550	5k & 18w	British victory
General Knyphausen	5,000	about 150k & w	American victory
Sir John Johnson, Jos. Brant	about 1,000		American victory
Gen. Benedict Arnold	1,700	48k, 145w	British victory
Major John Ross	700	40k & w, 30c	British victory

IN THE SOUTH—BRITISH

Commanders	Force	Casualties	Result
Gov. Lord Dunmore & Capts. Leslie & Fordyce	at least 200 out of 600	62k & w	American victory
Col. Donald McLeod	1,780	30k & w; 850c	American victory
Sir Peter Parker & Gen. Henry Clinton	1,780 9 warships	64k, 131w & 1 ship	Successfully held by Americans
Lt. Col. Archibald Campbell	3,500	3k & 10w	British victory
Major Gardiner	about 200		American victory
Colonel Boyd	700	40k & 70c	American victory

IN THE SOUTH—AMERICAN

Date	Battles	Commanders	Force	Casualties
Mar. 3	Briar Creek	General John Ashe & Col. Elbert	1,700	about 200k. 173c
June 20	Stone Ferry	Gens. Lincoln & Moultrie	1,200	146k & w
Sept. 16 to Oct. 18	Siege of Savannah	Gen. Benjamin Lincoln Adm. Comte d' Estaing	1,550 Amer. 3,500 French	total allied loss, 244k, 584w
1780				
Mar. 29 to May 12	Siege of Charleston	Gen. Benjamin Lincoln	about 5,000	92k, 148w & 4,650c
April 14	Monck's Corner	Gen. Isaac Huger	500	20k or w, 67c
May 6	Lenud's Ferry	Cols. Anthony White, Wm. Washington, Abraham Buford	about 350	41k & w, 67c
May 29	'Waxhaws	Col. Abraham Buford	about 400	113k, 203w & c
June 20	Ramseur's Mill	Col. Francis Locke	about 250 out of 400	about 150k & w
Aug. 1	Rocky Mount	Col. Thomas Sumter	600	14k & w
Aug. 6	Hanging Rock	Col. Thomas Sumter	800	12k & 41w
Aug. 16	Camden	Gens. Horatio Gates & Johann de Kalb	3,052	about 1,050 k, w & c
Aug. 18	Fishing Creek	Col. Thomas Sumter	700	150k & 330c
Oct. 7	King's Mountain	Cols. Campbell, Cleveland, McDowell, Sevier & Shelby	900	28k & 62w
Nov. 20	Blackstock's Plantation	Col. Thomas Sumter	about 1,000	3k & 5w
1781				
Jan. 17	Cowpens	Gen. Daniel Morgan	1,025	12k & 60w
Feb. 25	Haw River	Gen. Pickens & Col. Lee	at least 600	none
Mar. 6	Wetzell's Mill	Col. Otho Williams	600 to 700	about 20k & w
Mar. 15	Guilford Courthouse	Gen. Nathaniel Greene	4,400	78k & 183w
April 25	Petersburg (or Blandford)	Gens. Steuben & Muhlenburg	1,000	about 60k & w
April 25	Hobkirk's Hill	Gen. Nathanael Greene	1,551	19k & 115w
May 8–12	Fort Motte	Gen. Marion & Lt. Col. Lee	about 450	2k
May 22 to June 6	Siege of Augusta	Gen. Andrew Pickens, Cols. Henry Lee & Elijah Clark	over 1,600	16k & 35w
May 22 to June 19	Siege of Ninety-Six	Gens. Nathanael Greene & Andrew Pickens & Lt. Col. Lee	984 to about 1,500	57k & 70w
June 26	Spencer's Ordinary	Col. Richard Butler	570	9k, 14w & 32c
July 6	Greenspring Farm	Gens. Lafayette & Wayne	900	28k, 99w & 12c
Sept. 8	Eutaw Springs	Gen. Nathaniel Greene	2,200	139k, 375w
Sept. 12	Cane Creeke (Hillsboro Raid)	Gen. John Butler, Col. Maybin	400	25k, 90w & 10c
Sept. 28 to Oct. 19	Siege of Yorktown	Gen. Washington (Americans)	9,500 out of 11,133	23k & 65w
		Gen. Rochambeau (French)	7,800 soldiers plus 1,000 seamen	60k & 193w

Casualties are indicated by the following initials: k = killed; w = wounded; c = captured.

IN THE SOUTH—BRITISH

Commanders	Force	Casualties	Result
Lt. Col. Mark Prevost & Major Macpherson	900	5k & 11w	British victory
Lt. Col. John Maitland	900	26k & 103w	British victory
Gen. Augustine Prevost	3,200	40k & 63w	British victory
Gen. Sir Henry Clinton & Adm. Marriot Arbuthnot	about 14,000	76k & 189w	British victory
Lt. Col. Banastre Tarleton	650	3w	British victory
Lt. Col. Banastre Tarleton	150		British victory
Lt. Col. Banastre Tarleton	200 or 270	5k & 15w	British victory
Lt. Col. John Moore	about 700 out of 1,300	about 150k & w	American victory
Lt. Col. George Turnbull	about 180	about 12k & w	British victory
Major John Carden	about 500	200k & w	American victory
Gen. Charles Cornwallis & Lt. Cols. Tarleton & Rawdon	2,239	68k & 245w	British victory
Lt. Col. Banastre Tarleton	160	16k & w	British victory
Major Patrick Ferguson	about 1,000	157k, 163w, 698c	American victory
Lt. Col. Banastre Tarleton	270	about 50k & w	Draw or possible British victory
Lt. Col. Banastre Tarleton	1,100	100k, 229w, 829c	American victory
Colonel John Pyle	400	90k, over 250w	American victory
Lt. Cols. Tarleton & Webster	at least 1,200	at least 21k & w	Tactical draw
Gen. Charles Cornwallis	1,900	143k & 389w	British victory
Gens. Phillips & Arnold	2,500	about 70k & w	British victory
Lt. Col. Lord Rawdon	900	38k, 170w & 50c	British victory
Lt. Donald McPherson	175	175c	American victory
Lt. Cols. Thomas Browne & Grierson	630	52k & 334c	American victory
Lt. Col. John Cruger	550	27k & 58w	British victory
Lt. Col. John Simcoe	400	10k & 23w	Tactical draw
Gen. Charles Cornwallis	7,000	75k & w	British victory
Lt. Col. Alexander Stewart	2,000	85k, 351w & 440c	British victory
Cols. Fanning & McNeil	950	27k & 90w	American victory
General Charles Cornwallis	8,885	158 soldiers k & 326w, plus 44 seamen k & w, 7,247 soldiers & 840 seamen surrendered.	American victory

THE ANTAGONISTS

The Americans

CREATED IN JUNE 1775, the Continentals were the Regulars of the American Army. Often only Militia units given Continental numbers, these soldiers, unpaid, hungry and in rags, learned quickly and during Greene's Southern campaign, they fought many desperate actions where they proved themselves to be equal to British Regulars. After performing valiantly at Monmouth in June 1778, the Continentals never failed to give a good account of themselves against trained European troops.

The brigades of the Continental Army were usually formed of regiments of one State or closely related States. They varied in strength, some being equal to two regiments and others only having the strength of a single company, while many commanders organised their troops to suit themselves so that some company's were larger than other regiments. Often, they were so much below strength that one or more line company's would be dissolved and absorbed into remaining company's. Organised to resemble the British system, the paper-strength of a Continental regiment was 9 Companies, each of 8 officers and 68 men; eight being Regular infantry with the ninth an elite company of light infantry. The normal regimental officer strength was a colonel (or Lt.-Colonel), a major, an adjutant, a quartermaster, a surgeon and a paymaster with one captain and two lieutenants per company.

To create a standard-size unit, prevent an irregularly-formed line and make brigading easier, Von Steuben decreed that a battalion should consist of 200 men so that one regiment could form two or more battalions while a number of the 'company-size' regiments could form another. When possible militia were organised along the same lines although it was difficult because of the unpredictable number of militia who obeyed the call. The usual practice was to group militia in one large formation in the immediate vicinity of other militiamen from the same State.

Even in America during the Revolution, the formalities of 'Order of Battle' were observed in spite of difficulties because it was necessary to establish a priority for the various States. There was much bickering between the State brigades or the States within a brigade until 1778 when it was decided to form the brigades up in geographical order as the States presented their coast to the Atlantic Ocean. This meant that Massachusetts was on the left, then New Hampshire, Rhode Island, etc., all the way down to Georgia on the right flank. When the French were allied to the Americans they were given the left flank position as were the German mercenaries with the English army.

In the beginning, Washington's army wore their farm or other working clothes, and odd items of militia dress—all were festooned with ammunition belts and sword slings. The Militia discarded their scarlet military coats, replacing them with any civilian alternative (providing it was not red). They wore tricorne hat, breeches, waistcoat and gaiters, hose and shoes. In the earlier days only a few formations such as those from New Jersey and Washington's Virginia units wore blue, which was eventually adopted as the colour of the Continental army but it was not until 1781 that the majority of the soldiers were so dressed. Washington's general order of 2nd October 1779 changed the official colour of the Continental Army from brown to blue with differing faces for the various States: New Hampshire, Massachusetts, Rhode Island, and Connecticut wore white; New York and New Jersey wore buff; Pennsylvania, Delaware, Maryland and Virginia wore red and Georgia, North and South Carolinas blue with white edged buttonholes. Breeches and stockings were eventually replaced by long overalls and leggings made from deer skin, undyed linen or canvas duck, shaped to the leg and secured at the ankles with four buttons.

In a general order of 24th July 1776 Washington ordered his men to wear a type of hunting shirt saying '. . . it is a dress justly supposed to carry no small terror to the enemy who think that every such person is a complete marksman.' These hunting shirts were made of buckskin (deer leather) or home-spun linen, dyed various colours to distinguish between regiments—black, blue, brown, green, purple, white or yellow.

Both sides employed Ranger units who wore fringed buckskin shirts and leggings and, for parade purposes, short green jackets. Roger's Rangers were the most famous of these units, remaining loyal to the British.

The American infantryman was armed with several types of musket, ranging from the semi-fowling piece of the militiamen to the British Brown Bess which gunsmiths in 13 States were contracted to reproduce at the outbreak of hostilities. More of these weapons were obtained by raids and from British surrenders at Ticonderoga and Saratoga. Later in the war, French muskets were shipped into the country.

The American musket fired a round ball plus two buckshot, propelled by exploding powder measured into a paper cartridge. Shortage of ammunition sometimes forced the Americans to fire 'old nails and angular pieces of iron' as at Bunker Hill. The effective range of these weapons was between 80/100 yards.

Training drills taught the infantryman to achieve a rate of fire of one round every fifteen seconds in the heat of battle—a speed of firing which sent a hail of balls into the massed ranks of the enemy; firing was by the volley and careful aiming was not stressed.

The American rifleman was better armed than the British Light infantryman with his smooth-bore musket. The Kentucky rifle, with an effective range of 300–400 yards, was much superior to the British ·75 smooth bore Brown Bess muskets, causing Burgoyne to be of the opinion that his men were very indifferent musketeers so that they were encouraged to go in frontally with the bayonet—a disastrous tactic in the face of accurate fire.

Neither side used cavalry to any great extent, the Dragoon or the Light Dragoon were the only type of horse soldier, both trained to serve either mounted or dismounted. Their duties included scouting, raiding, patrolling, outposts and escort duties and rounding up deserters. On the march, the cavalry joined the light infantry as part of the advance guard; in battle they were often posted in the classic position on the wings of the line and, in a retreat, covered the rearguard.

The best-known American cavalry units were those of Henry Lee (Light Horse Harry); Armand's and that of Colonel William Washington. They were organised in four troops each of 50 men, operating as a Corps (each troop usually contained more than its authorised establishment). Irregular formations were usually 6 troops of 20 men each, on commandeered horses. Lee's was the only full-time cavalry unit—shortage of horses caused others to be temporarily organised.

In 1780 von Steuben mixed Dragoons with special infantry to form units known as Legions, and in December of that year Lee's Legion was 300 strong, fully uniformed and equipped and reported to be 'thoroughly disciplined scouts and raiders'. Washington's cavalry troop was in Greene's army of the South in December 1780 when it numbered only 90 men half equipped and uniformed but with a high morale. At Cowpens they defeated the famous Tarleton in a classic charge, after being checked by infantry fire.

Cavalry were armed with sabres and a large variety of pistols—both either captured or copied from the British.

The British and their Allies

Throughout the war at unit level the British fought with tenacity and sometimes dash although not always with intelligence. The British military writer and authority, Brigadier Peter Young is of the opinion that the British gave Washington and his men a better run for their money than the French, the Austrians or even the Prussians of the day could have done.

British infantry battalions of the American Revolutionary period consisted of ten company's—one grenadier company, one light infantry company and eight centre companies usually totalling about 400 although the 'paper' strength was 475 men. The Grenadier company, the elite of the regiment, was formed of the tallest and strongest men, who wore tall bearskin hats with large metal plates in front. They were stationed on the right of the regiment when drawn up in line, while on the other flank was the Light Company who operated as scouts or 'rangers'. In deference to their duties they had their coats cut short, wore small leather caps, and carried small cartridge boxes and hatchets. Rarely serving with their regiments, these two flank companies were generally brigaded with the flank companies of other units to form one elite Grenadier or Ranger battalion.

The British infantrymen wore a brick-red wool coat, with the facings on collar, lapels and cuffs of different colours to distinguish his regiment. His trousers and waistcoat were of white linen or wool; Highland regiments wore kilts of Black Watch tartan. He wore buckled black shoes partly covered by heavy linen gaiters. Around his body from his right shoulder a slung belt supported a 12-inch bayonet and his linen haversack containing food, with a tin canteen on top. Across his left shoulder he wore a black leather cartridge pouch, containing 18 to 30 rounds of ammunition, suspended by a whitened buff leather sling; on the flap of the box was a brass plate with a regimental number in a circle under a crown. At the start of the war the British infantryman wore the knapsack over his left shoulder with the haversack on the right; later he wore a painted canvas or white/brown goatskin knapsack square on his back. On his head he wore a black cocked hat. He was armed with the Brown Bess, a ·75 calibre flintlock musket weighing 15 lbs, and a 12-inch-bayonet.

Sergeants' coats were scarlet and their hats were bound in silver; they wore a scarlet sash with a stripe of the Regimental facing colour and carried short swords (hangers). Officers wore similar coats to sergeants, their hats were bound in metalic lace; a crimson net silk sash was worn around the waist and knotted on the left by officers of foot and on the right by officers of horse. Officers of Foot carried spontoons—long poles with an iron spearhead; their sword was suspended in a whitened buff leather belt from the right shoulder. Officers wore a rounded metal half-moon shaped gorget around the neck, tied with ribbon of the facing colour and engraved with the Royal Coat-of-Arms or the Regimental badge.

There were not many British cavalry in America—the 16th (Queens Own) Regiment of Light Dragoons fought at Brandywine, Paoli, and Monmouth. The 17th Regiment of Light Dragoons fought in the 1776 New York campaign and in the 1777 Philadelphia campaign; at Monmouth and Pound Ridge N.Y. One troop served in the Southern campaigns at Cowpens and Guilford Courthouse and were interned at Yorktown. Tarleton's Dragoons played a prominent part in the South—formed of 350 men operating in troops of 60/100, always serving as a unit and never split up.

Troopers in cavalry regiments serving in America wore the same basic uniform as the infantry with a helmet of metal or hard leather instead of the cocked hat. From their right shoulder they carried a straight Light Dragoon sabre and a wide belt from the left shoulder carried a carbine—a short 'Brown Bess'. Officers of Horse were armed with sabres of a better quality than the troopers and had a brace of flintlock pistols in saddle holsters. The troopers also carried pistols but they were of poorer quality.

German Mercenary Units in the British Service

German mercenaries were employed by the British Government during the Jacobite Rising of 1745 and had been used in other British conflicts so that in 1775 when hostilities broke out in America the Government hastened to employ the troops offered by the rulers of numerous small German States such as Hesse Cassel; Hesse Hanau; Brunswick; Waldeck; Ansbach-Beyreuth and Anhalt Zerbst. Hesse Cassel supplied 19 Regiments of Foot, an artillery corps, jagers and supporting troops; Brunswick provided 4 Foot Regiments, 1 grenadier battalion, supporting artillery and some Dragoons; Ansbach-Beyreuth sent 2 foot regiments, an artillery company and jagers whilst Hesse Hanau, Waldeck and Anhalt Zerbst sent a single foot regiment and supporting artillery.

The basic German regiment was a battalion with a Grenadier company; the Brunswick and Hesse Hanau regiments also had light companies. The paper strength of such a battalion was 21 officers, 60 sergeants and corporals, 22 drummers and 525 privates. The grenadier companies were regimented together in the British manner.

A Hessian fusilier or musketeer battalion had five companies—one grenadier and four fusiliers or musketeers. The 'paper' strength was 650 but casualties and desertions considerably depleted most German battalions.

The musketeer or private in the ranks wore a black cocked hat bound with white, with a coloured pom-pom; his coat was of blue wool with a left shoulder strap of a facing colour; light yellow or white waistcoat and breeches; a long pair of black linen gaiters over the breeches, stockings and shoe tops. The German grenadier wore a tall pewter-fronted mitre cap with a coloured (often yellow) bag in the cap part; it was surmounted by a coloured pom-pom.

The Germans wore their bayonets suspended from belts round their waists and carried short swords. A wide white or yellow sling-belt from the left shoulder held a large cartridge box with a metal coat-of-arms on its front; from the other shoulder was slung a knapsack of natural hair animal hide and a metal canteen. He was armed with a Prussian musket ·75 in calibre much like the British Brown Bess. German sergeants carried halberds (a long pole with an elaborate blade at the top); officers wore finer, profusely decorated uniforms and carried spontoons; wide sashes of different colours denoted their individual States.

The Germans had corps of jagers, raised from the hunters of the German forests, armed with heavy short rifles. They wore green coats with red facings and large cocked hats. Thinking them useful for the forests of America, the British Government hired large numbers of them but these units were at times no better than normal infantry battalions and they were not very successful because many of them were untrained peasants put into jager uniform.

Provincial Units in the British Army

A large provincial Corps was organised during the war, trained, armed and uniformed in much the same manner as Regular British regiments. A counterpart to the American Militia, these units were similarly irregular in numbers and tactical ability. Mostly organised into company's, occasionally they were formed into regiments when available in sufficient strength.

By December 1776 the Provincial Units wore uniforms of green coats faced with white, blue and green but for the 1778 campaign all Provincial units were clad in redcoats with different facings, in the same manner as the British regiments. Their weapons were older models of the Brown Bess carried by the Regulars. The best known of the Provincial units were the Rangers, often combined units of mounted and dismounted men—the Queen's Rangers had Dragoons, Riflemen, infantry and even a Highland company. Inspired by legends of the exploits of Roger's Rangers during the French and Indian Wars, the Rangers were the most effective Provincials and did valuable work in skirmishing, patrolling and carrying out raids on outposts, etc. They wore short green jackets and trousers and wore caps similar to the British Light Infantry head-gear; their leather belts and straps were darkened so that they were more often black but never white and many of them carried rifles and pistols instead of the Brown Bess.

Artillery

The employment of artillery was restricted by problems of land transportation in North America and guns were moved on inland waterways. The roads were so bad and the open ground so rough that movement of a train of artillery was a slow and laborious business requiring up to twelve horses to draw a light 12-pdr gun, with the trail of the gun attached to a wagon.

American artillery was organised on the European model, mainly on British lines and with British guns, acquired in large numbers when Burgoyne surrendered at Saratoga in October 1777. During the war French field pieces, obsolete in France because of the introduction of the Gribeauval system, were shipped to America. Later, French guns with French volunteers to serve them came into use.

In the attack on Trenton, Henry Knox a young Colonel of Artillery laboriously moved his cannon across the blizzard-swept River Delaware and dragged them along rutted roads covered in frost and snow to rake the streets of Trenton and prevent the Hessians, still affected by their Christmas celebrations, from assembling into battle formation.

In the campaigns in the South there was little scope for use of artillery as the fighting was a series of small local actions. In New England General Gates had 7,000 men including 2,700 of the Continental line and 22 guns, most of them of lower calibre than Burgoyne's.

Never very plentiful, American artillery pieces were individually placed to cover important fields of fire. With experience, American artillery became proficient and had probably attained parity with the British by late 1777.

The Continental Army started the war in 1775 with only the cannon, ammunition and gunpowder they could capture, plus a small quantity gathered together by the Colonial Militia, who had learned to cast cannon and make cartridges. Maximum effective artillery range was about 1,200 yards but with the relatively untrained gunners using old weapons and imperfect ammunition, only about 400 yards range was usual.

Washington's army had mostly bronze guns ranging from 3 pdr to 24 pdr with some $5\frac{1}{2}$ inch and 8 inch howitzers. He also had a few old iron siege guns of 18, 24 and 32 pdr. They used roundshot; grape and case shot; all rounds being carried in side boxes on each side of the gun carriage. They were drawn by horses or oxen; the Americans followed the British practice of hiring (or impressing) civilian drivers for their animals. Because these men were understandably reluctant to involve themselves in fighting, the last part of the approach into action was done by the crew members manhandling the guns with drag ropes attached to hooks on the axle, using a hand spike to manoeuvre the trail.

American artilleymen of the four numbered regiments of artillery did not have their own uniform until 1779 when they wore the tricorne hat and single breasted coats of blue, faced and lined with scarlet; white overalls with buttons down the side and black shoes were worn to give the general appearance much like that of the Royal Regiment of artillery in the British service.

The British guns were light 12 pdrs; light 6 pdrs; $5\frac{1}{2}$ inch howitzers; light 24 pdrs; light 3 pdrs (on Captain Congreave's carriage); mortars, medium and heavy 12 pdrs; medium and heavy 18 pdrs and heavy 32 pdrs. It was the type of carriage which caused the distinction between light and heavy guns; the light type were mounted on two large wheels which gave added mobility while heavy artillery was on a heavier carriage with four small wheels—the same type of mountings used on ships of the period. British Line infantry battalions had attached to them a light field piece manned by an N.C.O. and a dozen privates; sometimes these guns were grouped together to form a massed battery.

The 4th Battalion Royal Artillery Regiment served in America and wore blue coats faced with red; plain yellow lace and plain brass buttons; yellow bindings on hats. Burgoyne's army is said to have been well provided with artillery; his army of 4,646 had 45 guns—one to every 100 men. Most of them fell into American hands after the surrender at Saratoga and, during the course of the war, other guns were captured and used against their original owners.

THEIR STYLE OF FIGHTING

THE INFANTRY OF the American Revolution—American, British and German, fought well, employing tactics little inferior to those of the Napoleonic Wars while their skirmishing paved the way for Sir John Moore's famous Light Infantry that enabled Wellington to triumph in the Peninsular.

The battles of the American Revolution were contests of manoeuvre rather than of fighting ability largely because of the lack of artillery plus the rapid and accurate musketry which indicated the vital role to be played in European warfare by the rifleman and his rifle. So that the tactical importance of the American Revolution lies in its development of Light Infantry, and it was only in their light infantry that the armies of the American Revolution outclassed the European armies of their day. There were few light infantry on the battlefields of Europe up to the standard of those trained in the hard school of American frontier warfare, men like the Rangers or the American victors of King's Mountain. The British Light Companies were far superior to their European counterparts but their redcoats and white crossbelts were unsuitable for their role.

However, it was some time before it was realised the riflemen were light infantry but light infantry were not necessarily riflemen and, despite the efforts of Patrick Ferguson (whose rifle was well in advance of its time) the British did not immediately recognise the need for rifle units. At first they had one or two marksmen armed with a weapon to each battalion, but before the end of the war every British battalion in America had organised a Rifle Company.

British redcoats and their German allies were well drilled and formidable in a fire-fight on flat open ground—Washington did not like pitched battles under these circumstances! With the possible exception of the Prussians in Europe, the British line infantry at the time of the American Revolution could shoot faster and more accurately than anyone else; an American problem was to train their infantry to stand and trade volley-for-volley with the British, particularly when fighting in the settled regions along the coast where

European tactics prevailed. When fighting took place in the wilderness, where the frontiersmen were notably better in their use of cover, American irregular tactics were superior to those of the British and Hessians. Under the same conditions, the British Light Infantryman was probably superior to the American Line soldier. But the Revolution could not be won in the wilderness so American soldiers had to adapt themselves to the military thought and tactics of the day.

American troops generally fought in the accepted linear fashion of Europe; only in a few battles such as Lexington, Concord and King's Mountain did American troops fight from behind trees and stone walls while, in the first two actions, the British remained in tight formations on open ground. But, scorning the conventional European military code of war, the American did not like to fight shoulder-to-shoulder, loading and firing like automatons, rather they preferred to move forward as individuals. He did not regard retreat as shameful and, when things got hot, tended to withdraw as an individual, sometimes this lack of cohesion causing a rout. Taught in the hard school of Indian warfare, the woodsmen knew of the use of cover and night movement and the value of carefully aimed in dividual firing. Because this was a merciless style of fighting with no quarter, on the battlefield the American was more like an individual hunter than a soldier. The Americans did not like hand-to-hand fighting; this encouraged the British to depend upon cold steel and it is said that British soldiers prayed for rain so that they could attack with the bayonet without fear of having to take enemy fire.

The frontiersman with his long rifle has a very colourful and legendary place in the Revolution. His value was evidently appreciated from the very beginning because even before the Continental Congress appointed Washington to command, they resolved 'that six companies of expert riflemen be immediately raised in Pennsylvania, two in Maryland and two in Virginia'. They had much to commend them—American frontiersmen were said to be able to maintain a reasonably high rate of fire and also to be capable of re-loading whilst on the run; they excelled at skirmishing, sniping and scouting but suffered from the disadvantage of having a weapon that was slow to load and lacked a bayonet to meet British infantry in hand-to-hand fighting. Hard to pin down, the rifleman was respected by the British who are said to have sent a captive to England to demonstrate the efficiency of his weapon.

Seemingly Washington did not fully realise that his ill-disciplined frontiersmen with their great skill at arms were a great asset; attempts were constantly made to drill them in the rigid and formal European manner. Baron von Steuben mistakenly tried to counter the British on their own ground by endeavouring to instill rudiments of European drill into his American soldiers. It could be claimed that the finest hours of the American Revolutionary army were at Kings' Mountain, Freeman's Farm and Bemish Heights—victories of the frontiersmen in his deerskin battledress and the home spun New England farmer rather than the Continentals.

However, seemingly there were disadvantages connected with the frontiersman, who, although their marksmanship was impressive, were said to be '. . . otherwise quite useless except to pick off an occasional Regular who incautiously showed himself . . .' while their rowdy frontier habits soon provoked disciplinary problems. Except for special operations such as scouting, skirmishing, sniping, etc., the rifleman was said to be of little military value and that it was the man with the musket, who could pour out a high volume of fire and make his mark with the bayonet, who won or lost battles. Superior to musket men under certain conditions, riflemen were no match in a real battle because of their slow rate of fire and their lack of a bayonet. As time went on, the British began to recognise that, for these reasons, the American riflemen were not their greatest foes. The Americans themselves found that it was usually necessary to support outposts of riflemen with Regulars otherwise they were beaten-in and forced to retire.

When Morgan's Riflemen, flushed with their success over Burgoyne's army, attacked British light infantry under Colonel Abercrombie, he ordered his troops to charge them with the bayonet the moment they appeared before them. Not one rifleman out of four had time to fire and those that did had no time to re-load, so that Abercrombie's light infantry dispersed them at once and drove them for miles over the country. It is said that they never attacked, or even looked at British light infantry again, without a Regular force to support them.

The British and their German mercenaries used *linear tactics*, with a line of battle consisting of two or three ranks drawn up shoulder-to-shoulder with another rank of 'file closers' about six paces behind, to replace casualties. Linear formation took a long time to form and, once formed, was difficult to manoeuvre so that it had two choices:

1. Either it had to advance to win by disciplined delivery of sustained fire or
2. It was defeated by its inability to take such fire from a steadier enemy.

In meticulous alignment the attackers, their bayonets gleaming, moved forward with their officers shouting and beating at them with their spontoons to make them hold their fire until they were within 50 yards. The principle of this style of warfare was to take the enemy's fire, withstand the losses and then put in your own volley at such close range that every shot told.

Volley firing was done in a formal fashion with all loading and firing carried out on the word of command; there was little or no aiming and the volley was delivered directly ahead or slightly obliquely as commanded. The intention was to lay down a pattern or field of fire by firing rapidly but not necessarily accurately. Training was designed to improve speed of loading and firing, so as to severely decimate the attackers (who were reasonably safe until inside 100 yards range) before they were close enough to be effective. Because well-trained soldiers could fire a shot every 15 seconds, it was generally assumed that the closely-formed and compact attacking mass would have to take two volleys before closing.

Following the twelve set motions of the Manual of Arms for the Brown Bess, an indifferent soldier could fire two rounds in a minute, whilst a well-trained veteran could get off as many as five in the same time. The first round was always more effective because it had been loaded in a careful, leisurely and supervised manner, being properly rammed down before the bayonet was fixed; the long bayonet made it difficult to ram down subsequent rounds.

It was reasonable to expect a bullet to strike home at 80 yards or even at 100 yards but a man wounded at 150 yards or over could consider himself very unlucky; very few people were ever killed or wounded at more than 200 yards range, although the ounce lead ball fired from a musket with a calibre of nearly $\frac{3}{4}$inch diameter had sufficient velocity to be effective up to 300 yards.

THE COMMANDERS

THE AMERICAN MILITARY LEADERS during the Revolution were a mixed bunch with Washington himself standing head and shoulders above them. Realistic and practical, George Washington was a better strategist than any commander on the British side and although some of his generals (notably Greene), displayed talent, there was no credible alternative as Commander-in-Chief. The British military writer Peter Young says 'Washington was far from being one of the great generals of all time although a man of sterling character and unquestioned courage, with a very respectable fighting record in the Seven Years War and an almost proverbial integrity. That he won a war with such an army is not the least proof of his military talent.'*

General Horatio Gates, a former British Army officer with experience in the French and Indian Wars, was commissioned as a General in the American Army at the outbreak of the Revolution and was at one time seriously considered as a replacement for Washington. He won a reputation at Saratoga in 1777 but lost it in 1780 when his poorly trained and equipped force of 4,000 men was routed by 2,300 British troops in a night action at Camden, where Gates fled the battlefield.

General Nathaniel Greene. With Washington at Trenton, he fought at Brandywine and Germantown. Commander of the Southern Army from 1780, he handled the Battle of Guilford Courthouse in an intelligent manner.

Major-General Marquis de Lafayette, a glamorous figure of the American Revolution, he was wounded at Brandywine and was at Valley Forge with Washington.

Colonel Henry Lee. His mounted raids were legendary and he and his legion played a big part in the American Revolution.

General Artemas Ward, veteran of the French and Indian Wars, was in nominal command of the American at Bunker Hill.

Colonel William Prescott successfully organised the defences of

*'George Washington's Army'—Peter Young. (Osprey, England 1973).

Breeds Hill (Bunker Hill) where his handling of an Irregular group of Revolutionaries in the face of a superior body of trained soldiers was noteworthy.

Colonel Daniel Morgan. Led a corps of riflemen who played a prominent part throughout the Revolution. His victory over Tarleton at the Battle of Cowpens is a classic of the American Revolution.

Major-General Baron von Steuben, this German professional soldier had fought in the campaigns of Frederick the Great and is most noteworthy for his training ability, turning Washington's farmers and woodsmen into a respectable army.

General Anthony Wayne. Although known as 'Mad Anthony' he was a very cool soldier in battle and fought well at Brandywine, Germantown, Stoney Point and Monmouth.

General Benedict Arnold. This arch villain of American history was an extremely brave and dashing leader who showed up well at Bemis Heights and at numerous other battles.

Among the best (or the most effective) American commanders were the guerilla leaders like *Francis Marion* the 'Swamp Fox'. *General Thomas Sumter* the leader of a band of South Carolina guerilla fighters, Sumter displayed many talents learned in frontier fighting in a series of minor operations against Tarleton, while *Ethan Allen* successfully led the Green Mountain Boys.

Regimental officers were elected by their men among whom they moved more or less as equals; some of them had command experience in the Militia whilst a few had served during the campaigns of the Seven Years War.

With the possible exception of Cornwallis, the rest of the British commanders—Burgoyne, Clinton and Howe—were a mediocre group, although it has been claimed that these Britsh leaders were experienced and competent professionals whose leadership was badly affected by the low standards of regimental and company officers. There was no shortage of officers in the British army but most of them were poorly trained and with little military aptitude. Victories usually occurred because the troops they commanded were sufficiently well-trained and disciplined to perform creditably in spite of poor leadership at company and regimental level.

The young British commander, Banastre Tarleton earned a reputation during the war as an almost legendary leader of light cavalry, being unmatched by any other man on either side for dash, daring and swift manoeuvres. Disliked by his fellow officers for his brashness, he was thought to lack the military maturity for anything but a partisan raid or a cavalry charge.

RE-FIGHTING THE BATTLES OF THE WAR OF THE AMERICAN REVOLUTION

TO RE-FIGHT ON THE wargames table the best known battles of the American Revolution—King's Mountain, Bennington, Brandywine, Monmouth, Cowpens, Guilford Courthouse and others—is a most attractive proposition. It can be practically achieved with a considerable degree of realism by following the advice offered on these pages. The conflicts of this famous war are highly suitable for reproduction on table-top battlefields because they involve small numbers so that realistic scaling-down is possible. For example, Brandywine in 1777 was one of the biggest battles of the war yet Washington's army totalled only 10,500 men and at Camden in 1780 Gates with 3,000 was driven from the field by Cornwallis with 2,400, yet by the standards of this war it was a major battle.

If the battle is to bear anything more than a resemblance in name to the original then the table-top armies should be in accurate proportion to the original forces, both in numbers and types. The terrain must closely resemble (both in scale and appearance) the area over which the original conflict raged; the size and topographical features of the American battlefields lend themselves to reproduction on the wargames table, being both suitable in area and frontage for this purpose. Each battle seems to have tactical and human interests that make it worthy of simulation.

The table-top reconstruction of historical battles must conform as closely as possible to the events that occurred on the battlefield itself —it is pointless to construct a terrain resembling that of Cowpens on which a host of infantry, cavalry and artillery mill around in a manner that in no way resembles the original conflict. Every aspect of the battle has to be considered in its correct context and in chronological order, only then can it be realistically simulated although its outcome may be affected by the Military Possibilities of the day or the fluctuations and fortunes of war.

By re-fighting these battles of the American Revolution, the wargamer will gain an understanding of the problems of the commanders in the field and an insight into the military thinking of the time, because using the formations and weapons of the day, he is refighting each battle in the correct technical manner. The purpose of this book is to suggest practical methods of simulation that will produce an accurate, realistic and enjoyable reconstruction of the original battle.

Of course, the most authentic battle will result if the troops perform *precisely* the same manoeuvres as they did on the day, take the same percentage of losses and achieve the same success or failure. But this is not a wargame, it is an historical exercise, ideal as a demonstration of what occurred during the real life battle. Otherwise, one can follow the original course of events with sufficient leeway and a little stretch of imagination so as to allow a reversed result, but if too many liberties are taken the battle will become merely a wargame and not a reconstruction.

When reconstructing any historical battle it is essential that the tactical limitations of the troops of that day are taken into account, otherwise the incongruous situation could occur where an undisciplined group of frontiersmen perform formal military evolutions far more 'European' in their style than those carried out by their British opponents. This can happen simply because the American commander might be more conversant with the formal tactics of that period than is the 'General' leading the British.

All troops on both sides must conform to their known standard 'Style of Fighting'. For example at Cowpens Morgan required his militia to loose off two rounds of musketry; after that he was quite prepared for them to melt away. If, on the wargames table those militia are allowed to stand and fight stoutly in the manner of the Maryland Line or the British Guards then it is quite unreal. Hindsight should not allow table-top commanders to perform tactical manoeuvres unknown to their counterparts on the historic field.

Reading accounts of a campaign and its battles enable the wargamer to appreciate its tactics and objectives, such as Greene during the Southern campaigns displaying strategical sense by fighting not to win but to cause Cornwallis such losses as to diminish his future prospects of victory. Knowledge of a commander's historical intentions allows the wargamer to better understand his tactical plans.

The wargamer should analyse the battle, consider those moves that led to victory or defeat, while seeking possible alternatives to the historical trend of events. Called Military Possibilities, such alternatives can be severally defined in the case of Guilford Courthouse:

Allow the American second line to hold and not be pushed back.

Webster's men hold instead of withdrawing, or withdraw but do not rally.

Allow the British Guards to beat the Maryland Regiments and to turn and fight off the cavalry rear attack.

How would all this have affected the battle? You must find out on the wargames table!

The wargamer can transfer himself into the mind and body of George Washington, of Cornwallis, Clinton and Howe, or as the ill-fated Burgoyne, attempt to reverse the situation at Saratoga. He can don a buckskin hunting coat and become Dan Morgan.

Military Possibilities

Over and above the conventional rules that control the action, there are certain factors frequently mentioned in this book which aid in achieving an accurate and realistic reconstruction. Known as Military Possibilities, they are controlled and logical alternative courses of action that, taken at the appropriate time in the battle, might have completely changed its course and reversed its ending. Sometimes the course of action indicated by Military Possiblities results in a more reasonable and credible result than occurred on the historic field of conflict! Essentially courses of action agreed upon and followed by both wargamers, Military Possibilities must not serve as excuses for diverting the course of history merely 'to see what happens'.

Because Luck plays a big part in war, some Military Possibilities depend upon it in the shape of the throw of a dice or the turn of a Chance Card. The simplest means of simulating the fluctuations of fortune, Chance Cards introduce pleasant and unpleasant factors to materially affect a battle, perhaps even its result. Posing tactical, psychological or physiological eventualities, the instruction on each card have to be practically applied by the commander drawing the card. For example, the card may say that there has been a torrential rainstorm which has so affected the road as to delay the arrival of artillery or that a column has gone the wrong way and will not arrive until three game-moves have elapsed. It is an indication of the potential authenticity of Chance Cards that both those eventualities happened to Napoleon at Waterloo—obviously he drew the wrong cards! Each campaign can have a set of cards designed to cover topographical and tactical factors peculiar to that operation—their trend and number depends largely upon the ingenuity of the imaginative wargamer.

The instructions on Chance Cards form the human element that can affect the purely tactical aspects of a Military Possibility. For example, at King's Mountain in the Fall of 1780, only small numbers of Mountain Men rendezvous at Sycamore Shoals on the Watauga because of bad weather; then, during their march towards Gilbert Town, the deep snow on the crest of the mountains can cause more to drop out. Ferguson learned of the Rebels approach through his agents—a Chance Card can prevent this happening. Halting his retreat on October 6th at King's Mountain, Ferguson anticipated being reinforced but the illness of Tarleton, Hanger and Cornwallis delayed or prevented support being sent—a Chance Card can restore these British Commanders to their normal good health! The rocky nature of the countryside around King's Mountain itself can be made even rockier by an umpire with a twisted sense of humour, while the frustrating instructions on a Chance Card makes the Mountain Men come up in small groups instead of as a cohesive force.

So much for Chance Cards, that optional accessory that brings to wargames a fascinating yet irrational aspect strangely similar to some of the 'goofs' that bedevil real soldiers and armies.

According to the choice of the wargamer the use of Military Possibilities can radically alter the historical course of a battle or be resticted to some relatively minor aspect, both possibilities can cause interesting tactical twists and may well influence the eventual outcome of the engagement. There are numerous Military Possibilities suggested in the reconstructed battles of the American Revolution; such as at Cowpens on January 17th 1781 when the militia of Morgan's first line might be made to characteristically break instead of retreating steadily to make a complete circle of the American force and hit the British in the flank. This had a relatively decisive effect on the result of the battle; had it not occurred then Tarleton might well have won.

Another Military Possibility might allow General Charles Lee, who dragged his feet almost decisively in the early stages of the Battle of Monmouth on June 28th 1778, to be up-rated so that the true worth of the troops under his command was displayed against the three opposing brigades of crack British troops. Both sides claimed Monmouth as a victory but it could well have been an undisputed American triumph. Military Possibilities make such things possible; the interest and colour they bring to the table-top battlefield depend largely upon the imagination of the wargamer.

The Classification of Commanders

When two opposing forces are equal in strength, morale, equipment, positions, manoeuverability, etc., then victory almost certainly goes to the force with the best commander. It was a marked feature of the American Revolution that many strong forces were defeated by numerically weaker armies lucky enough to be commanded by a man with superior tactical ability or who was capable of inspiring his men to exceptional heights. On the other hand, he might have lacked both of those qualities and just happened to be better than his opposing commander! With hindsight, the history of the American Revolution reveals the occasions when one commander was 'Above Average' while his opponent was only 'Average' or 'Below Average'; for the sake of authenticity and realism, this fact has to be reflected upon the wargames table.

With the possible exception of George Washington, there was no historically outstanding leader present at any of the battles described in this book, so that the rating of commanders must be by 'local' comparison. The reconstruction notes for each battle include a suggested classification of commanders but the 'Above Average' rating given to Greene at Guilford Courthouse is merely by comparison with Cornwallis, his opposite number and does not imply that Greene was a commander of the calibre of Napoleon, or even Washington himself. Greene did enough to earn his superior classification although, because of the disparity in the size of their armies, that classification will probably put victory, on the table-top, completely beyond the reach of Cornwallis who can, at best, only be classified as an 'Average' commander. The performance of the British Colonel Webster at Guilford Courthouse entitles him to be considered 'Above Average' along with Colonel William Washington, whose cavalry performed admirably.

The classification of a commander affects the course of a table-top battle through the behaviour of his troops, since men commanded by an 'Above Average' commander must be given a higher standard of morale and better fighting qualities—represented by adding 1 (or any pre-decided score) to all dice affecting morale or fighting qualities. Conversely, a 'Below Average' commander would necessitate deductions from dice scores; while the troops of an 'Average' commander would remain unaffected.

An 'Above Average' commander could be given greater scope in movement and, within certain limits, he is allowed to undertake tactical movements that are historically outside the period. This means that the wargamer is allowed to be 'himself' in that he can employ up-to-date tactics with which the commander of 1776 was not conversant. With certain reservations, the 'Average' commader has the ability to alter characteristic fighting patterns, controlling and manoeuvering his force with permitted discretions over and above their laid-down 'Style of Fighting'. The 'Below Average' commander is allowed no such privileges and must control and move his force in strict accordance with their 'Style of Fighting'.

The classification of commanders can be simulated by the manner in which he gives his orders; for example, the rules compel the commander to write his orders at the beginning of the game, telling the 2nd Maryland Continental Line that they are to 'move forward, take the hill and hold it'. But the prevailing tactical situation might make rubbish of these instructions because the wargamer in charge of the 2nd Maryland is not going to send them blindly forward when the top of the hill is occupied by a battery of artillery and about three times as many red-coated British infantry! So the orders have to be altered and the higher the commander's classification the quicker he can change his orders.

In effect, this means that the 2nd Maryland, having moved forward say two game-moves out onto the table-top terrain, will need a galloper sent to them with a message cancelling their original instructions and giving them fresh orders. The 'Below Average' commander takes two moves to weigh up the situation, write his message and instruct his messenger; the 'Average' commander takes only one move to carry out the same actions whilst the 'Above Average' commander whips off a message without any delay, displaying all that brilliant quick thinking which has made him what he is!

In those cases, where one of their commanders is classified as 'Below Average', American readers should not feel slighted because this classification is solely based on the commander's tactical ability and does not reflect upon his courage. It is an odd historical fact that most 'Below Average' commanders were men of great personal courage, such as Custer at the Little Big Horn. Perhaps such men compensated for marked tactical deficiencies by displaying bravery to the point of foolhardiness.

Brandywine—11 September 1777

In an effort to thwart Howe's advance towards Philadelphia, Washington positioned his 11,000 troops along Brandywine Creek, which was a barrier in so far as it could only be crossed by its numerous fords. With most of the artillery, Greene's division and Wayne's brigade were just south of Chadd's ford with Maxwell's 800 Light Infantry deployed on the enemy side of the creek. The American

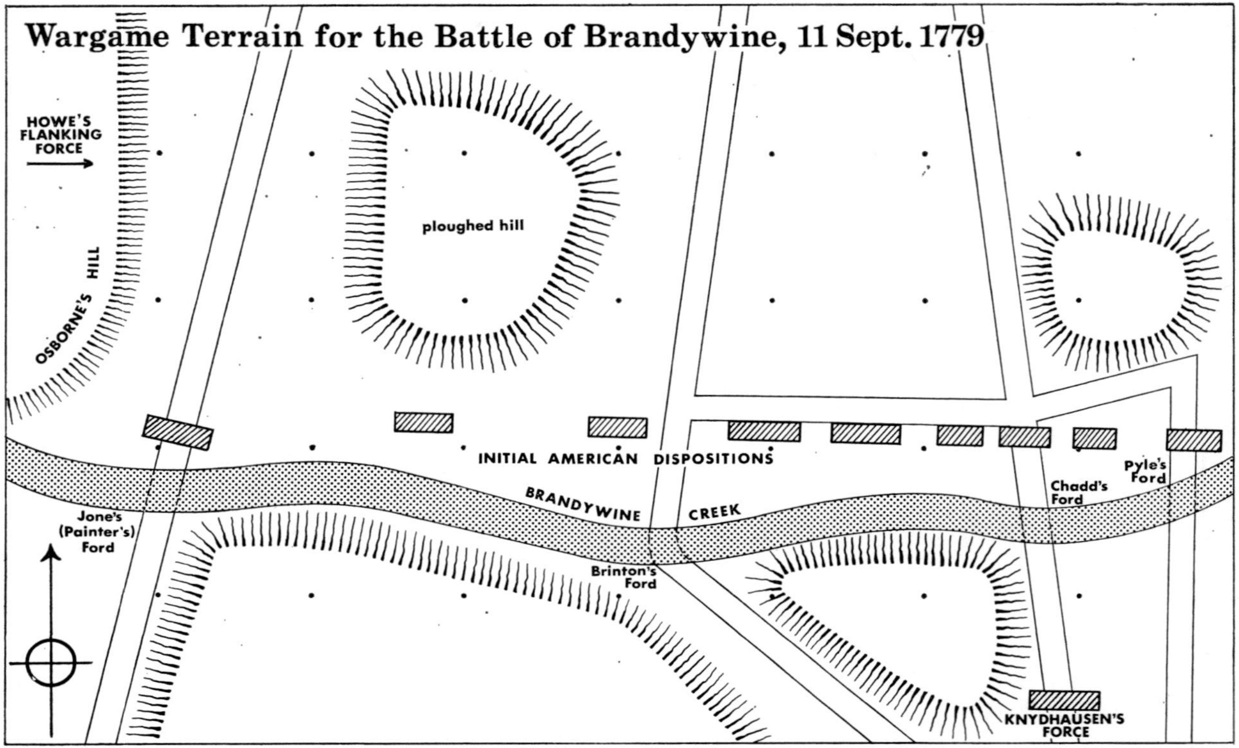
Wargame Terrain for the Battle of Brandywine, 11 Sept. 1779
HOWE'S FLANKING FORCE
OSBORNE'S HILL
ploughed hill
INITIAL AMERICAN DISPOSITIONS
BRANDYWINE CREEK
Jone's (Painter's) Ford
Brinton's Ford
Chadd's Ford
Pyle's Ford
KNYDHAUSEN'S FORCE

left was guarded by Armstrong's Pennsylvania Militia at Pyle's ford half a mile downstream. Sullivan's division were responsible for all fords north of Chadd's ford. East of the creek on the right Alexander's division were positioned to support either Sullivan or Greene, together with Steven's division who had to support the defenders of Chadd's ford. Washington's right was screened by all available light cavalry and mounted patrols operated a few miles west of the Brandywine.

Commanding some 12,500 men, Howe's plan was for Knyphausen with 5,000 men to make a secondary attack against Chadd's ford while Howe and Cornwallis led 7,500 in a turning movement round Washington's right.

The secondary attack pushed Maxwell back over the ford and by mid-morning Knyphausen was in position along the creek, his artillery exchanging shots with Porter's American guns on the far bank.

Although the British lack of effort at forcing a crossing at Chadd's made Washington suspicious that Howe was attempting to turn his flank (the subject of numerous contradictory messages in reports received throughout the morning and early afternoon), the American commander took no positive action. Then at 2 o'clock, the American cavalry screen on the right reported two enemy brigades near Osborne Hill two miles north of Sullivan's flank. Ordering Sullivan to march immediately to meet the new threat, Washington also sent Alexander and Steven's formations hurrying towards Birmingham whilst the American Commander stayed at Chadd's ford with Greene's division and the brigades of Wayne and Atwell. At 4.30 there were the sound of cannons in Washington's rear and exchanges of musketry volleys and, at the same time, British artillery fire quickened at Chadd's ford. At the head of Greene's division, Washington started across country towards the sound of firing.

Alexander and Steven had occupied good defensive ground on a ploughed hill south-west of Birmingham and had almost completed their deployment when Sullivan arrived and took over command. There was some re-positioning as Sullivan's formation fitted into place and finally the regiments of Hazen, Dayton and Ogden on Sullivan's left and the other divisions on his right held off the first British attacks with the American artillery showing up prominently.

Notwithstanding, when Washington arrived at 5.30 the British Guards were beginning to break Sullivan's left and the rest of the line was sagging, being restored on the right by Weedon's brigade when they arrived at six o'clock. Earlier this flank had been gallantly held by Colonel Marshal's small regiment of 170 men supported by the guns of Woodford's brigade of Steven's division. The infantry and artillery survivors dropped back through Weedon's brigade as they took up their position. Leaving Sullivan to carry on the action, Washington and Lafayette rode up and down the line, the latter being wounded. Soon Sullivan's wing was beaten and were streaming back along the road to Chester.

Knyphausen had pushed across Chadd's ford after pouring a heavy artillery fire on Wayne and Maxwell before the 71st Highland Regiment followed by the Queen's Rangers and the 4th British Regiment together with the rest of the command poured across the creek. The American left fell back and four guns were lost and then the British came up with the American line with four guns drawn up on a hill in front of a forest with their right wing resting on the Chester road. Armstrong's Militia withdrew from Pyle's without seeing any action and, with the Americans retreating on all sides Knyphausen and Cornwallis made contact at about 7 p.m. Howe did not pursue as he considered his troops were too tired. The British had 577 casualties and the Americans lost 1,250 men and 11 guns.

Reconstructing the Battle as a Wargame

Surprise, that most difficult facet of wargaming, has to be simulated here to represent the turning movement around Washington's right carried out by Howe and Cornwallis. Should the wargamer playing Washington be familiar with the course of the battle, then the best that can be achieved is for him to be compelled to lay his forces out in their historical initial dispositions and then to have his future courses of action strictly controlled by a series of time-limits on the movement of those forces which opposed the British flank attack. A more realistic method is for a wargamer unfamiliar with this specific battle to be chosen to act as Washington and to receive, at the appropriate time, the information and messages actually received by Washington in 1777. He will then act upon them in accordance with his own discretion and judgement. If this reconstruction is to be more than a nominal one, the American commander must not be allowed to place his troops in any positions that will radically alter the course of the battle—any reversal of its result must be achieved by table-top generalship and the sheer fighting ability of the American troops.

Despite being surprised, the Americans fought well and certainly merit a classification of 'Average'. The British, like the professionals they were, displayed discipline and competence and should also be classified as 'Average'—the reward for their superior tactics will be reflected in the morale-rating of their opponents. The British Guards were elite troops and both their fighting and morale classification should be subject to a bonus.

General Howe the British Commander completely outwitted Washington in his tactical planning of this battle and therefore merits a rating of 'Above Average'. His immediate subordinates Cornwallis and Knyphausen both displayed 'Average' ability on the field of battle. Washington the American Commander, according to his biographer, '. . . conducted the Brandywine operation as if he had been in a daze'. And notwithstanding his high reputation he cannot be classified in this case as anything but 'Below Average'. His immediate subordinates General Sullivan, held by many people to be responsible for this defeat, must similarly be 'Below Average'. Other American Commanders, particularly Wayne, merit the classification of 'Average'.

With things going so much their way, the British troops were obviously in a high state of morale throughout. Until actually aware that they had been outflanked, American morale was good but in all fighting subsequent to the outflanking contact on the American right (including that against Knyphausen at Chadd's Ford) the American morale must be subject to penalities on those occasions when it is checked.

The style of fighting of both sides was 'in period'—the British shoulder-to-shoulder in tight formations and the Americans only able to match them when aided by such terrain features as woods and shrubbery. In the open, the American formations must definitely fight at a disadvantage against their better trained and disciplined opponents.

The terrain played a vital part in the American defeat through its series of fords which made possible Howe's outflanking movement and provided a reason for Washington's initial dispositions. Hills and woodland play their part both in concealing the British outflanking march and the American defensive positions once contact is made. If this simulation is to be accurate then there is little room for changes brought about by Military Possibilities. By allowing Washington to react earlier than historically occurred, the whole trend of the battle will be altered as Howe will not reap the reward of his superior tactical thinking. On the other hand had Washington displayed even slightly better generalship he could have contained Cornwallis with minimum forces long enough to crush Knyphausen at Chadd's Ford. Taking this further, a Commander such as Napoleon might well have gone on from there to use his interior lines and defeat the other enemy wing. There might be a logically permissible Military Possibility in the fact that Knyphausen's secondary attack was prematurely executed with insufficient vigor to 'fix' the bulk of Washington's force so that their attention was occupied whilst the turning movement was executed.

Freeman's Farm—19 September 1777

DESERTED BY HIS Indian scouts, General Burgoyne the British Commander lacked information of the enemy, who were well entrenched on Bemis Heights, so that he advanced through the broken wooded terrain in three uncoordinated columns Burgoyne himself with Hamilton commanded in the centre, General Fraser was on the right and General von Riedesel on the left. By noon the centre column had reached the clearing in which stood Freeman's Farm, with Riedesel's column about a mile and a half to the east on the River Road, and Fraser in the woods to the west.

Suddenly, to turkey-calls signals, Morgan's riflemen crept within range and picked off every officer of the advance guard around the farm, causing such panic that the remainder of the guard rushed back on their main body without firing a shot. The riflemen chased after them but were dispersed by volleys from the British infantry. Burgoyne advanced his column to form along the northern edge of the clearing of Freeman's Farm with the 20th, 62nd and 21st regiments arranged from left to right and the 9th regiment in reserve (only the battalion companies of these four regiments were involved as their light and grenadier flank companies were with Fraser). Morgan's riflemen and Dearborn's Light Infantry took up firing positions along the southern edge of the clearing, supported by seven other American regiments of Continental line brought down from Bemis Heights by General Benedict Arnold.

On the British right, Fraser's force (grenadier and light companies, a battalion of Loyalists, a Regular Battalion, some guns and a few Indians) turned back Arnold's attempt at outflanking the centre column. Then they remained relatively inactive on the heights for the rest of the day.

For 3-4 hours the three British regiments stood shoulder-to-shoulder in the European manner under a constant heavy fire which repeatedly sent them back to the shelter of the woods. When the superior numbers of Americans charged into the clearing in pursuit they were driven back by British discipline, bayonets and the four light guns, two on each flank. American reinforcements allowed them to extend their line and force the three British regiments to thin out to prevent being overlapped on their right; nevertheless they repeatedly went forward in bayonet attacks. Suffering heavy losses,

Wargame Terrain for the Battle of FREEMAN'S FARM

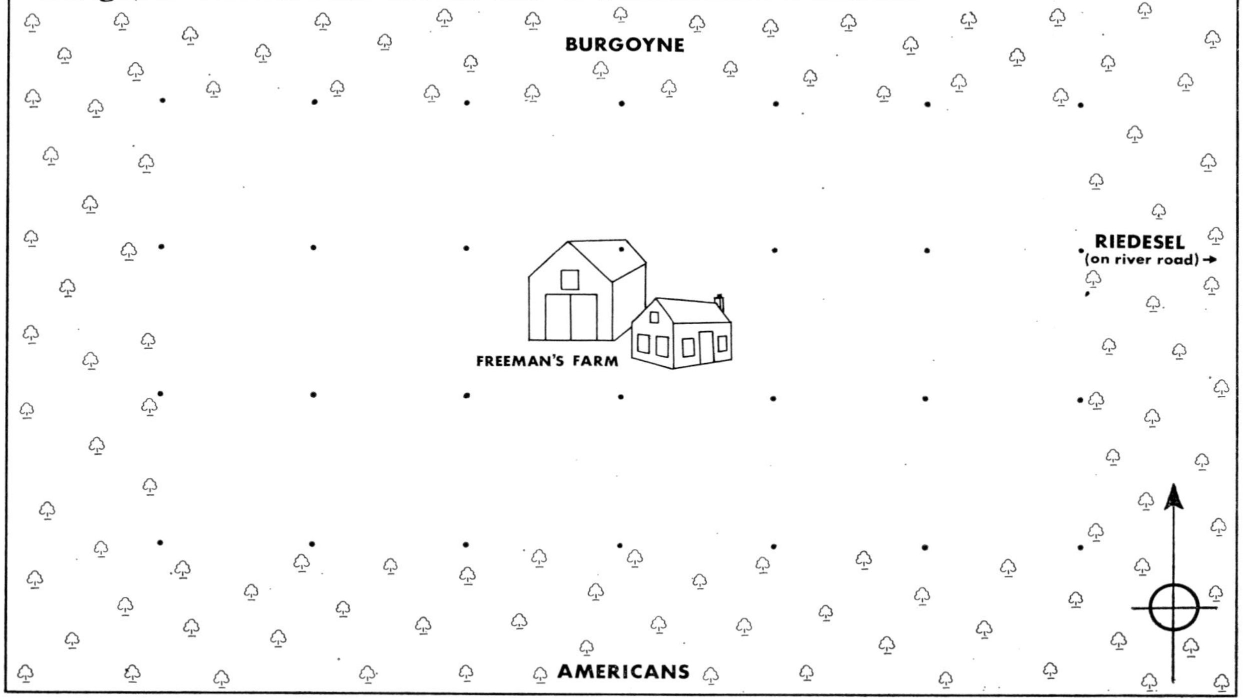

the centre column became exhausted and in a desperate situation when the Germans came to their aid.

Hearing the fire-fight, Riedesel had sent four guns to support the centre column and, receiving orders from Burgoyne at 5 o'clock, left some 800 men to protect the baggage along the river road and moved out with 500 infantry and two guns. Cheering and beating their drums, the small force advanced on the American right flank, where there were no patrols covering the regiments attacking the centre column. Arnold was back at Bemis Heights trying to persuade the lethargic General Gates to send reinforcements. Lacking leadership the Americans began to fall back and, as darkness was falling, Burgoyne with Riedesel's reinforcements launched a counter-attack which caused the Americans to withdraw. Burgoyne camped on the battlefield and, had he continued his attack on the following day, he would probably have defeated Gates.

Burgoyne lost 600 out of the 800 men in three British regiments; 44 % were casualties. American losses were 319.

Reconstructing the Battle as a Wargame

The fact that Burgoyne the British Commander was unaware of the whereabouts of the American Army can be simulated by erecting a curtain or screen across that half of the table which lies the American position and ruling that Burgoyne's force moves forward in a blind condition until reaching the Freeman's Farm clearing (only the British half of this area being visible); the screen is then lifted and Burgoyne's advance guard finds themselves confronted by Morgan's Riflemen, as in real-life.

Because the battle took place almost exclusively between Burgoyne's centre column with aid from Riedesel's column coming at a later stage in the battle, it is only necessary for Burgoyne's force and part of Frasers to actually be on the wargames table. Riedesel's force will be moving 'on the map' and will only come on the wargames table at the appropriate time. To simulate the surprise impact of Riedesel on the American right flank, the latter should be unaware of the existence of this flanking column, only becoming conscious of it when they actually arrive on the table.

All the troops involved in this battle display good military qualities, with Morgan's Riflemen completely filling their appropriate role. The other American Regiments fought well, particularly when aided by covered terrain but (as in almost every engagement of this war), did not show up so well when involved in a formed fire-fight or melee contact. This must be reflected here by the superior numbers of Americans inevitably forcing back their British opponents by sheer weight of fire but when following-up the British tactical withdrawals, the subsequent melee rules must be so framed as to allow the well disciplined bayonet tipped close formations of the British Regulars to have an advantage.

The very character of the American soldier and the nature of his military training and formations inevitably made his morale-state more variable than that of the highly professional British soldiers. This was personified at Freeman's Farm by the manner in which the American force faltered when under pressure and bereft of General Arnold's command. To accurately reconstruct this battle, Arnold must be absent from the wargames table onwards from a specified time, and 'local rules' designed to reflect this.

RATING OF COMMANDERS:

British—	Burgoyne	—Average
	Hamilton	—Average
	Fraser	—Average
	von Riedesel	—Above Average
	Phillips	—Average
American—	Gates	—should not be given any actual *battlefield* control as Arnold was in practical command—but should control distribution of troops to battlefield, acting as a Below-Average commander.
	Arnold	—Above average
	Morgan	—Average
	Poor	—Average

Although relatively simple, the nature of this terrain was vital to the course of the battle. The clearing in which stood Freeman's Farm was the focal point for most of the action; the surrounding woods first providing cover for Morgan's Riflemen and then allowing the Americans to fight in their preferred style by utilising its cover whilst providing a haven for the intermittent tactical withdrawals of the British when forced back by the fire of their numerically superior American opponents. Similarly, the broken wooded terrain allowed Riedesel's reinforcements to arrive unseen on the American right flank.

Apart from the normal eventualities of war when the British could have broken in the face of their numerically superior opponents, the outcome of this battle could have been affected at three points. 1. Had General Arnold not been back in the camp with Gates apparently at the moment when Burgoyne was reinforced, the American morale and standard of fighting would probably have remained high. Had

Riedesel not arrived then Burgoyne would have been hard pressed to even hold his ground let alone avoid defeat. At the same time, if the Americans had had patrols on their right flank so as to warn them of the arrival of Riedesel then the German leader's shock-impact might well have been neutralised, with subsequent effect upon the eventual result of the engagement.

Germantown—4 October 1777

Howe split his army into two parts—the garrison of Philadelphia under Cornwallis and a main body of 9,000 men five miles north in and around Germantown. Reinforced to 8,000 Continentals and 3,000 Militia, Washington planned a dawn attack on the Germantown force and at 7 p.m. Washington's columns had started forward, his ill-shod troops marching slowly on the rough roads. The American leader planned for Greene with three large divisions (forming two-thirds of the total American force) to make a wide envelopment through Lucken's Mill against the British right; Sullivan was to advance along the Shippack Road with Alexander's division in reserve; Maryland and New Jersey Militia were operating on the American north flank and Pennsylvania Militia on the south.

In spite of heavy fog that reduced visibility to 30 yards, British outposts reported the American advance at 3 a.m. With Greene not yet in position on his left, Sullivan sent Wayne to cover that flank and two other regiments were sent to protect the right before Sullivan's entire force moved forward, pushing back British pickets who had been reinforced by light infantry. In spite of the poor visibility, Colonel Musgrave's 40th Regiment came forward and carried out a stout delaying action, counter-attacking Sullivan's men and stoutly defending the numerous lines of fencing as they slowly withdrew. Finally Musgrave's force came to the Chew House which they occupied with six weak companies (120 men). The divisions of Sullivan and Wayne were past before Musgrave's force opened fire from the Chew House but Maxwell's divisions was held up for more than half an hour.

Greene's column (who had lost Stephen's division in the fog) pushed on through the fog to come up on the British right at Luckens Mill. Then, finding themselves in danger of being outflanked the American formations veered south towards the Market House where the British had superior forces and were able to more than hold their own.

The Centre column pushed on and Washington, seeing Stephen's men on Wayne's left, assumed that Greene's command was in

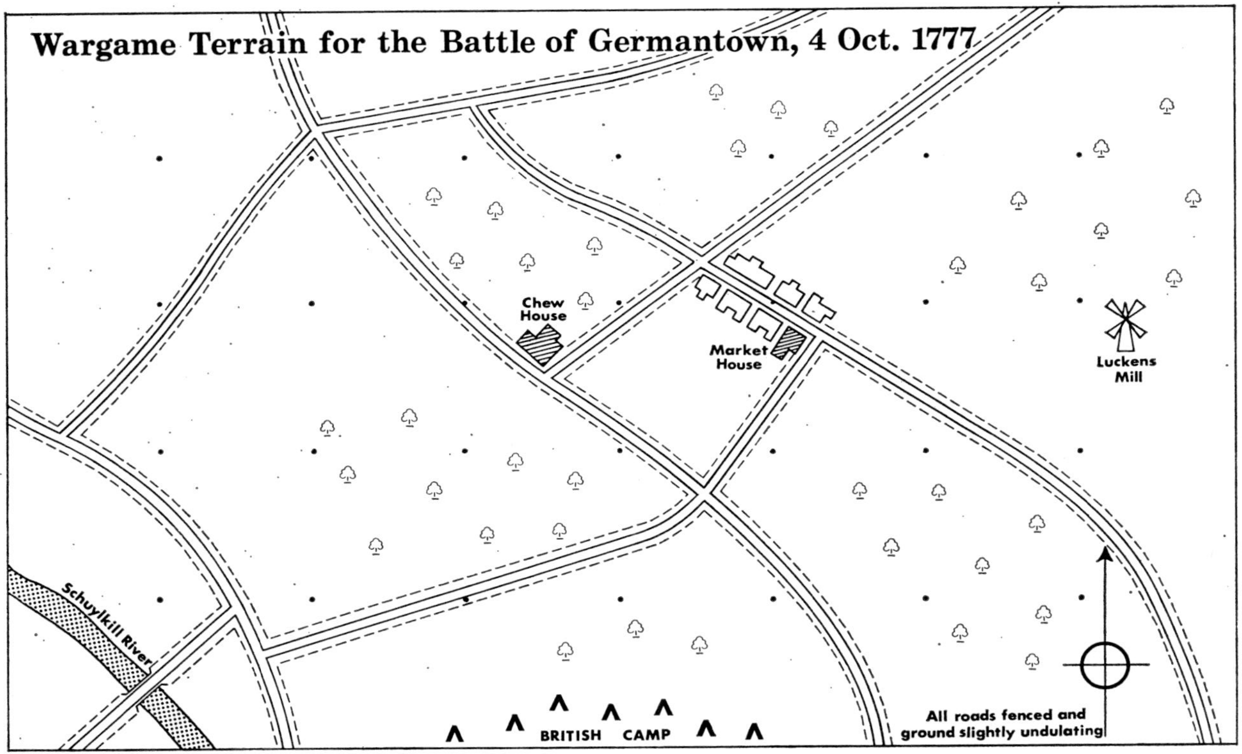
Wargame Terrain for the Battle of Germantown, 4 Oct. 1777
Chew House
Market House
Luckens Mill
Schuylkill River
BRITISH CAMP
All roads fenced and ground slightly undulating

position on that flank and prepared to carry out his planned attack. On the extreme left the Militia took no part in the battle, but Armstrong's Militia on the other flank along the Schuylkill River were in contact with the British. Suddenly on the left, where Wayne's men were extending Sullivan's line, there was confused firing and shouting —Wayne's brigade, coming upon Stephen's division in the fog had mistaken them for the enemy and fired upon them. This caused a degree of confusion that sent men running back through the fog frantic with fear and, in spite of Washington and his Staff attempting to rally the disturbed Americans, fear turned to panic and the entire American Army were suddenly transformed into a horde of fugitives. Although the worst of the panic ended quite quickly, Washington had little alternative but to order a general retreat. The British did not pursue vigorously although Cornwallis brought three fresh battalions from Philadelphia and followed Greene for some five miles without bringing him to action.

The Americans lost nearly 1,000 killed, wounded and prisoners; the British casualties were 550.

Reconstructing the Battle as a Wargame

So much of the blame for the disastrous outcome of this battle was blamed by the Americans on the fog—and as this natural phenomena is intensely difficult to simulate on the wargames table, a realistic reproduction might be difficult. On the other hand, if the fog resulted in the American downfall then equally it allowed them to make an abnormal unopposed contact with Howe's strong force. It is not unreasonable to state that anything but a British victory would have been unlikely because Washington (no doubt with his Trenton success looming large in his mind) had formed the highly ambitious plan of making a night march over unfamiliar roads with his army of 11,000 'amateur' soldiers divided into four columns with flanks seven miles apart. Even if this march has succeeded, on arrival his weary force had to assault a formed army of 9,000 professional soldiers defending a good terrain on a three mile front.

At the opening of this reconstruction the British Commander, unaware of numbers and nearness of the enemy, will have his main body back in camp, with a reasonable number of British pickets forward on the table. As soon as they make contact, they are reinforced by light infantry followed by Musgrave's 40th Regiment. Then with the American formations moving forward as historically specified, the British will debouch from their camp and dispose themselves in correspondingly opposing positions.

No wargamer worthy of his salt is going to voluntarily fire upon one of his own regiments (although the author can remember a well known wargamer who, after much careful manoeuvering, charged one of his own regiments) so Wayne's firing upon Stephens must occur as a result of fluctuation of fortune. There is no better way of representing this on the wargames table than by throwing a dice, so a 'local' rule should be made (for *both* sides) that whenever ANY formation comes within range of another (even if it be one of its own) then one dice is thrown for each formation and a score of say 1 means that one regiment fires upon the other. Should this fire be received via a flank, then the force receiving it must immediately check their morale with heavy 'shock' penalties. Should they break and run, then the rules should be so formulated as to make a chain-reaction possible among adjacent formations.

In what fighting there was, the combat qualities of both British and American were seemingly normal—the eventual American rout was a matter of morale rather than fighting qualities. If the highly creditable display of Musgrave's 40th Regiment is to be simulated in the face of greatly superior numbers then they must be classified as 'Above Average' troops (i.e. elite).

Like Inkerman, another battle fought in a fog, this was a 'soldiers' battle with little control being exercised by any commanders on either side. The sole exception is Colonel Musgrave of the 40th Regiment who should be classified as 'Above Average'. The remainder of the commanders on both sides can be 'Average'.

The obvious Military Possibility that might have altered the course of events was for Wayne's brigade not to have fired upon Stephens and, should this not occur in our reconstruction, then the wargame will fight itself out in a normal way.

The terrain itself seemingly played little part in this battle with the exception of the Chew House and its surrounding garden.

Bemis Heights—7 October 1777

After the Battle of Freeman's Farm, Burgoyne dug-in, building the Balcarres Redoubt in the Freeman's Farm area, the Breymann Redoubt 500 yards north and erected stockades, and three great redoubts were started on high ground overlooking the area. Seeking information of the American positions on Bemis Heights, Burgoyne planned a reconnaissance-in-force by 1,500 of his remaining 5,000 Regulars plus 600 auxiliaries under Captain Fraser. At about 10,30 a.m. Burgoyne's force moved forward in three columns whilst Fraser's auxiliaries vanished into the wooded hills on their right

Wargame Terrain for the Battle of BEMIS HEIGHTS, 7 Oct. 1777

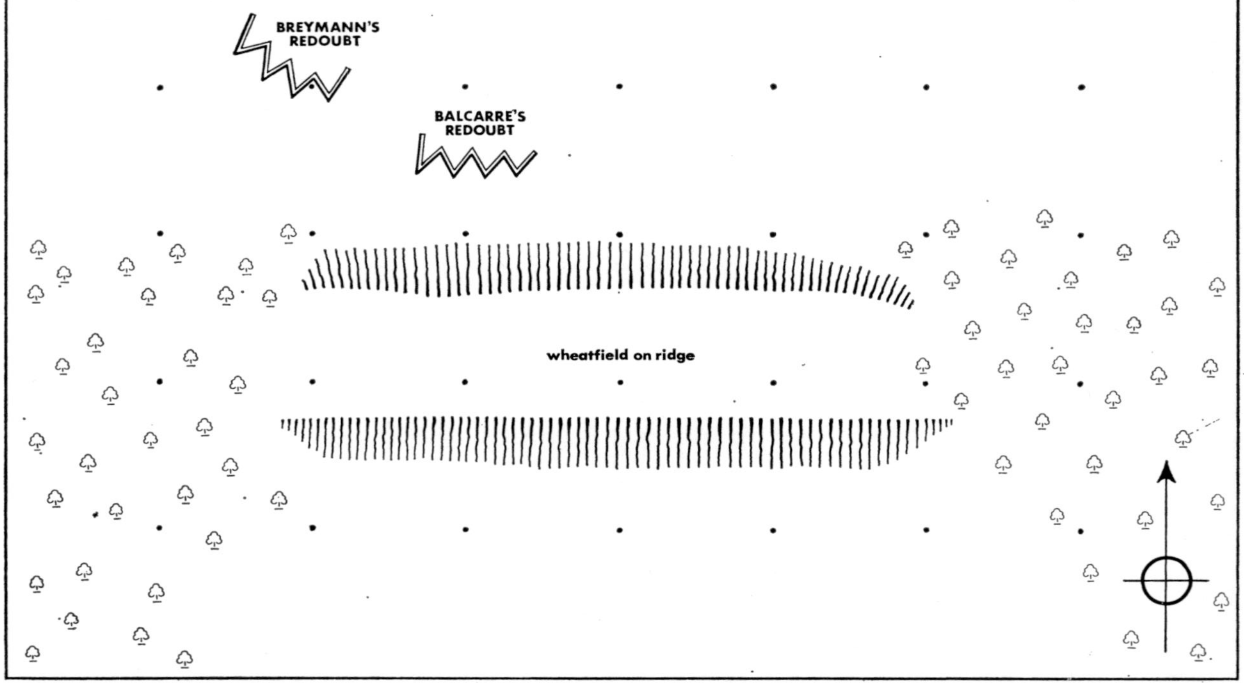

flank. When ⅓ mile on, Burgoyne's 1,500 men took up a good position in a wheat field on a gentle rise north of Mill Creek which gave an excellent frontal field of fire for their ten guns. The right flank of the line was held by Balcarres with his light infantry; Riedesel was in the centre with a composite group of Brunswickers and the 24th Regiment; and Major Acland with the British Grenadiers held the left of the line. Both flanks rested against thick woods.

But the American soldiers did not play the game of war in the European manner, possessing a rooted dislike of charges across open ground into the mouths of waiting cannon and formed ranks of disciplined Regulars. They were much better fighters in the woods and there were plenty of trees and cover on the approaches to both flanks of Burgoyne's line.

Subsequently, Morgan's Riflemen were sent to work around the high ground to turn Burgoyne's right whilst Poor took his brigade of 800 veterans of the New Hampshire Continental Regiments of Cilley, Hale and Scammell; von Cortlandt's and Livingston's New York Militia and the Connecticut Militia of Cook and Lattimer to attack the elevation held by Acland's force. Out-numbered 2/3 to 1, the Grenadiers opened fire with musket and cannon and then Acland ordered them to fix bayonets and charge. As they swarmed down the hill the Continental Line first delivered a murderous volley that stopped them dead in their tracks and then, whooping and cheering, charged in with the bayonet to sweep over their position and send them reeling back leaving Acland wounded on the ground.

On the other flank Morgan had routed Fraser's auxiliaries in the woods and turning the British line, hit Balcarres in flank and rear. As the British Light Infantry were changing front to meet this threat they took a devastating volley from Dearborn's Light Infantry that sent them reeling back in rout. Before they had gone far Balcarres rallied them but was unable to prevent withdrawals from the field.

Simultaneously, General Arnold (relieved of his command and without authority) dashed at the British centre at the head of three regiments of Learned's Brigade. With both flanks uncovered, the German troops in the centre fought on, reinforced by detachments from the Rhetz and Hesse-Hanau Regiments. They drove off the first attack but, remorselessly encouraged by Arnold, overwhelming American strength assailed them on three sides causing them to retire back to the Balcarres Redoubt in good order. General Fraser attempted a delaying action with the 24th Regiment but his conspicuous figure on a grey horse made him the target for snipers and soon he was shot from the saddle.

In the past the Americans would have been content with such a victory but driven forward by Arnold, they made a general attack upon the British entrenchments. The Light Infantry and other British survivors of the action repulsed the American attackers after they had got through the abatis. Then Arnold led them and Learned's brigade in an attack that cleared the stockade between the Balcarres and Breymann Redoubts, before racing to the other flank to lead four regiments (Wesson; Livingston; Morgan's and Brook's Mass. Regiment) in a successful assault on the redoubt. Here Breymann himself was shot dead by one of his own men after he had personally sabred four of his own soldiers 'to encourage the others'.

The Americans lost only 150 men while Burgoyne had some 600 casualties and lost all ten guns in a defeat that forced him to ask for terms, leading to his subsequent surrender.

Reconstruction of the Battle as a Wargame

As the British advance was a reconnaissance-in-force designed to seek out the American positions and their numbers, the table-top battlefield is best set up with the British in their battle position with flanks resting on thick woods. This cover should be fairly extensive but sufficiently spaced to allow action to take place within its confines. This allows reconstruction of Morgan's defeat of Captain Fraser's auxiliaries and Poor's advance through the woods to emerge upon the surprised Acland. On the other flank Arnold similarly sets upon Balcarres. In both cases the surprise element plus superior numbers caused the British flanks to withdraw, leaving their centre (mainly consisting of German troops) outflanked and, after a temporary resistance, forced to withdraw.

The American success owed much to the stimulating leadership of General Arnold which must be reproduced on the table-top by giving a combat-bonus to those units he is personally leading. Thus Arnold will be an 'Above Average' General, as will Morgan and Poor.

On the British side, Burgoyne will be 'Below Average'; Acland and Balcarres will be 'Average', while General Fraser's stimulating presence will rate him 'Above Average'. Falling victim to the continued efforts of American sharpshooters, Fraser should be removed from the table-top battlefield at the appropriate moment as his continued presence is historically inaccurate and might well result in an anachronistic British victory!

The British force were heavily outnumbered at all times—for wargames purposes this could be a convenient two-one. This, coupled with Arnold's inspired leadership, was the reason why the

Americans (although they all fought in first-class fashion) were able to so severely defeat the trained and disciplined British and German Regulars.

The course of events will inevitably lead to fluctuations in morale, although both sides began the battle at a high morale level. In the case of the British this wavered but the Americans remained high.

This battle is almost classical in the contrasting styles of fighting which marked this war, the British at Bemis Heights fought shoulder-to-shoulder in their European style while the Americans characteristically took advantage of the wooded approach to the British flanks. Then having used that advantage to shatter the British flanks, the Americans eventually succeeded in driving back the centre in a characteristic attack of the period; even so, they were repulsed at their first attempt.

The type of terrain is vital to the course of the battle—without the woods on the British flanks the American approach would have been handled in the same manner as it was in many other battles of this war.

Only by the widest stretch of the imagination can Military Possibilities be discovered in this battle. Perhaps if Acland's Guards had held on the British left (as they might conceivably expect to) then things could have been different, although Balcarres was thrown back on the right so exposing the flank of the British centre. If General Fraser had not been brought down he *might* have inspired his force to hold, but this is conjecture that might eventually rest on a straight comparison between the stimulating effects of Arnold upon the Americans and Fraser on the British.

Monmouth—28 June 1778

Slowly retreating from Philadelphia over bad roads and in very hot weather conditions, Clinton's force reached Monmouth with the pursuing Americans attempting to bring them to battle. At 4 a.m. on the 28th, Clinton sent Knyphausen north with a portion of the army and the baggage train, remaining behind with Cornwallis and the rear guard. This consisted of 14 battalions and the 16th Light Dragoons—the 3rd, 4th and 5th British infantry brigades, the 1st and 2nd British Grenadier battalions, all the German Grenadiers, the British Guards, the 1st and 2nd British light infantry battalions; Simcoe's Tory Queen's Rangers.

Washington sent General Lee forward with an advance guard of 5,000 men and 12 guns to attack the rear guard, although Lee was ill suited for such a task, as he did not believe that the Americans

Wargame Terrain for the Battle of Monmouth, 28 June 1778

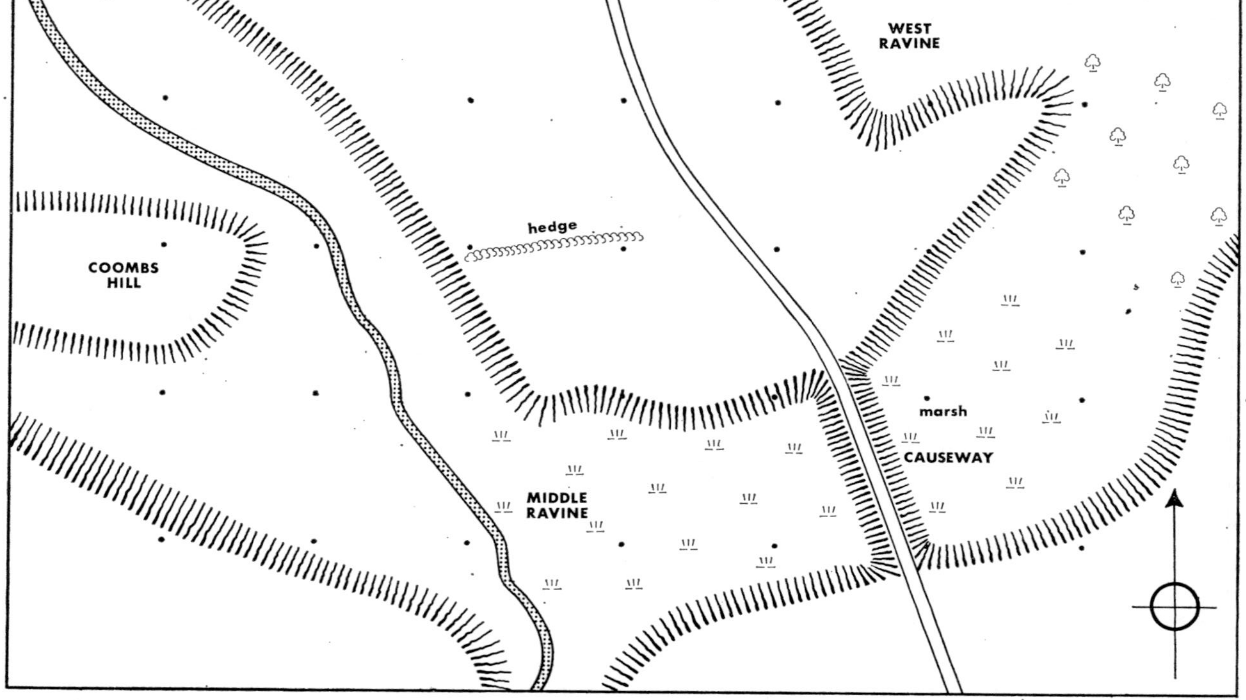

could stand against British bayonets. The first Americans to make contact were Colonel Grayson with his own Virginia Regiment, Scott's Brigade, part of Varnum's Brigade and 4 guns (a total of 600 men) and Dickinson's 1,000 strong New Jersey Militia. The remainder of the American order of march was Colonel Butler with 200 men; Colonel Jackson with another 200, then part of Woodford's Brigade and Varnum's Brigade (each of 600 men and 2 guns); Wayne with 1,000 men and 4 guns; Scott with 1,400 men and 4 guns and Maxwell with 1,000 men and 2 guns. Lee ordered Lafayette to take three regiments of Wayne's command and some artillery against the enemy's left flank. But his manoeuvres were mistaken by his fellow commanders for a withdrawal and first Scott and Maxwell and then Grayson and Jackson dropped back to start a general and confused retreat with some units in good order and others completely disorganised. This retreat was encouraged by the gleam of British bayonets as Clinton, fearful for the safety of his army and its long baggage train, turned at bay.

Washington, riding in advance of the main body, was astounded to come upon his retreating army and, after a stormy confrontation with Lee, took over personal command. Hastily he deployed his main body to the right and left of the road forming a new line to the rear and then went forward to halt the retreat, an incredibly difficult task with the routing troops of a relatively new army in the face of the enemy.

Meanwhile, Clinton had decided to bring on a general engagement and re-grouped his forces in two lines—from left to right the British Grenadiers, Guards and Light Infantry; the 4th Brigade, 3rd Brigade and the Hessian Grenadiers behind them, all advancing steadily on the dusty heels of the retreating Americans. Partly behind a hedgerow on a ridge between the Middle and West ravines, Washington formed a line from the troops of Livingstone, Varnum, Stewart and Ramsay; the centre commanded by Wayne, with Greene on the right and Stirling on the left and with Lafayette in charge of the reserve. Four guns came into action to support this forward line and Greene got another four guns onto Comb's Hill, under the protection of Woodford's Brigade, from where they delivered an effective enfilade fire on the British left and centre. The American position was strengthened by a swamp east of the ridge which presented a serious obstacle to the advancing British.

For the next hour the British Light Infantry and the 42nd Regiment (Black Watch) with some field guns, endeavoured to envelop the American left on the high ground, but the personal leadership of Washington, Steuben and Stirling caused the ground to be held. Then a counter-attack through heavy woods of the American left by the

1st and 3rd New Hampshire and the 1st Virginia Continental Line drive back the British right and relieved the threat. Cornwallis then led a massive attack of the Coldstream Guards, another Guards battalion, English and Hessian Grenadiers, the Light Infantry detachment and the 37th and 44th Foot against Greene's right wing. This attack was brought to a halt and driven back, mainly because of the flanking fire from the guns of Comb's Hill. Simultaneously, Wayne's centre was hit by a series of frontal attacks—the first by Light Infantry, Dragoons and Grenadiers was shattered by close-range volley fire; a second attempt met the same fate. Then an attack made by Colonel Monkton after forming-up less than 500 feet from the American line was also halted, Monkton being killed.

With the temperature nearing 100 degrees and both his flanks threatened, Wayne withdrew in good order, his men falling back on the main position—which was far too strong for the British to hope to carry. A counter-attack ordered by Washington came to nothing when the advancing regiments fell apart soon after starting through heat, fatigue and failing light.

The similarly exhausted British troops were withdrawn out of range and the battle subsided. Clinton slipped away undetected at night and caught up with Knyphausen on their way to New York.

Both sides claimed victory in the longest action of the entire war. The Americans lost 356 and the British 358 men.

Reconstruction of the Battle as a Wargame

If this battle begins on the table-top at that same point of time as it did in real-life then it is unlikely to unfold in an historically accurate manner because:

(a) it is remotely possible that Lee *might* not retreat, which would alter the course of the battle and eliminate its cardinal features

(b) assuming that the reconstruction takes a realistic course, the subsequent morale-state of the retreating Americans would be so adversely affected as to make it virtually impossible for Washington to rally them.

So, it is best to begin the battle at the stage when Washington takes over command of a 'Below Average' force led by 'Average' leaders, with 'local' rules allowing graduated increases in the morale-status of the Americans as commander/units achieve isolated successes. In this way, both commanders and their units can build themselves up to combat morale statues of 'Average' or even 'Above Average'.

The prime feature of this battle was the conditions of intense heat under which it was fought. This can be simulated by another 'local' rule which makes deductions from the fighting capabilities of both sides. Thus, the Americans will deduct one from their combat scoring after each *alternate* game-move but the British/Germans fighting on alien soil and handicapped to a greater extent by their unsuitable uniforms and weight of equipment carried, will deduct one at the conclusion of *each* game-move.

Clinton and Cornwallis, the British commanders, can be rated as 'Average' commanders together with their subordinate leaders. Washington will commence the battle as an 'Above Average' commander, as will Steuben and Stirling. The remainder of the American commanders, such as Wayne, Maxwell, Lafayette, Greene, etc., will commence the battle with an 'Average' classification but 'local' rules will give facilities for up-grading to 'Above Average' in accordance with their progressive successes.

This extremely hard-fought battle is claimed to be the point at which the American soldier 'found himself' and realised that he could hold his own (aided by his characteristic style of fighting and intelligent use of his native terrain) with the trained and disciplined Regulars.

Initially low, the morale-status of the American formations must, by 'local' rules, be allowed to rise rapidly. As always in this war, closely integrated with discipline and training the British morale was first-class and is subject to the usual combat-fluctuations.

The terrain played an important part in the course of the battle, particularly from the American point of view as their resurgence owed much to intelligent use of the hedge, Coomb's Hill and the swamp—all these topographical features must be reproduced on the wargames table and suitably embodied in the action.

This dourly fought drawn battle, claimed by both sides as a victory, reveals no marked errors, failings or inadequacies after the initial American retreat had been stayed. Therefore there is little scope for realistic Military Possibilities except for the major aspect which could alter the entire course of the battle—if Washington had failed to rally his commanders and their men.

Camden—16 August 1780

Advancing towards Camden, General Gates led a force of about 4,100, made up of Stevens Virginia Militia (700); Kalb's Delware and Maryland Continentals (900); Armand's Legion (120 mounted and foot); Portersfield's Virginia Light Infantry (100); North Carolina Militia (1,800); 70 Volunteer horsemen and 6 guns. With this motley force of sick and exhausted men, Gates incredibly decided to make a

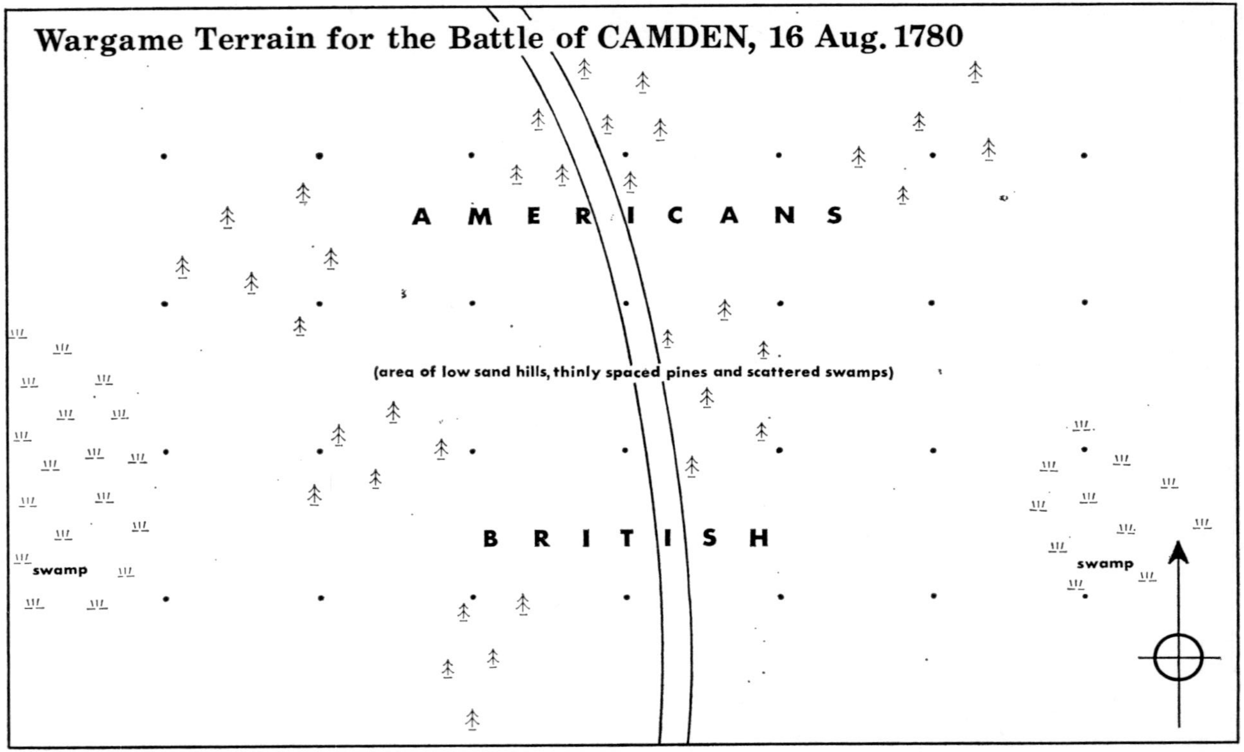

Wargame Terrain for the Battle of CAMDEN, 16 Aug. 1780
AMERICANS
(area of low sand hills, thinly spaced pines and scattered swamps)
BRITISH
swamp
swamp

night march and attack Cornwallis at daybreak.

By an uncanny coincidence Cornwallis had left Camden at 10 p.m. and was marching along the same road towards Gates with a view to also attacking the Americans at daybreak! At 2.30 a.m. the two armies met at Parker's Old Field in Gum Swamp, and the British 'point' (20 mounted and 20 dismounted dragoons of the British Legion) charged and drove Armand's advance guard back in confusion, but were themselves halted when American flank patrols closed in. After some desultory firing, both sides began to take up their positions for battle in a sandy area of widely spaced pines narrowed by dense swamps. Gates was on slightly higher ground but his flanks would be 'in the air' if he had to withdraw from the narrowest point in the field whilst Cornwallis, handicapped by the narrow front and lack of depth for his reserves, was less than a mile forward of Saunder's Creek, a considerable stream some 200 feet wide that could turn defeat into annihilation.

The British force, made up largely of well seasoned Regulars, deployed in a line perpendicular to the road with their Light Infantry on the extreme right against the swamps; the 23rd (282 men) and the 33rd (283 men) in a wing under Colonel Webster. Then came the Volunteers of Ireland (287 men); the Infantry of the British Legion (150); the North Carolina Tories (250 men) extended left to another swamp. Colonel Bryan's North Carolina Tory Volunteers (300 men) were in echelon to the left rear of this flank; the whole wing under the command of Lord Rawdon. The 71st Highlanders (237 men) were to the rear astride the road with Tarleton's cavalry (100 sabres) to the rear of the road behind the Highlanders, formed in column through lack of space.

Parallel to the British the American line from East to West were the Militia of Stevens on the flank, with Armand's Legion to their rear; Caswell's North Carolina Militia towards the centre of the line; Gist's 2nd Maryland Brigade west of the road—the whole forming the right wing under General Kalb. Smallwood's 1st Maryland Brigade was astride the road to the rear as a reserve and the six guns of the 1st Virginia Artillery were posted in front of the American centre near the road. The big fault in this formation lay in the fact that Gates had his Militia on the left opposite the British Regulars and had kept half his Regulars in reserve.

Forming up the armies occupied them until dawn when cannon began firing on both sides and soon the battlefield was wreathed in a heavy fog.

Situated some 600 yards behind the battle line, Gates gave no orders and was apparently content to await attack. This soon came when the British right wing, firing and cheering, advanced rapidly forward which threw Stevens Virginia Militia into such a panic that they dropped their loaded arms and fled. They were instantly followed by the North Carolina Militia then by others, some after firing two or three rounds. Most of the Militia, forming at least two-thirds of the American Army, fled without firing a shot.

However, the American Regulars in reserve reformed their ranks after being broken by the fleeing Militia and held their ground against repeated attacks. The British right did not pursue the militia but wheeled to roll up the exposed American right flank. Amid the reduced visibility caused by fog, dust and smoke, the American Regulars tried to reform but were unable to close the 300 yard gap between brigades. Cornwallis now turned Webster's Regulars against the front of the reserve brigade who were twice driven back and twice rallied before the Marylanders were driven from the field. Hand-to-hand combat was now taking place between the attacking British and the other brigade and General Kalb was killed. The exposed flank of the American right were attacked by a part of the Legion cavalry and Tarleton, returning from pursuit of the left wing, hit them in the rear. Only about 60 men of the American Army retreated as a unit, others scattered individually and in small groups. The British pursued until exhausted and the pitiful remnants of the American Army reassembled at Charlotte before moving on to Hillsboro where only 700 men remained. Gates army was so completely annihilated that there are few records of casualties—Fortescue claims about 1,000 Americans were killed and wounded and over 1,000 captured. The British lost 324 men in what was later called 'The most disastrous defeat ever inflicted on an American army'.

Reconstruction of the Battle as a Wargame

This was a classic 'encounter' battle so that the wargame should commence with both forces set up on the table-top battlefield in their historical dispositions. Subsequently, the high quality of British commanders and their troops against the low-quality militia and the outnumbered first-class American Continental Line Regulars, will most probably cause this battle to unfold exactly as it did in real-life.

The classification of the commanders of both sides is clearly reflected by the course of the battle. The magnitude of the British victory, supported by contemporary reports of the battle, indicate that Cornwallis's tactics and handling of the action were first-class and that his subordinates—Rawdon, Webster and Tarleton—supported him superbly. All deserve the classification of 'Above Average'. General Gates, the American commander, appears to have exercised no control over the battle and his initial dispositions were faulty in the extreme—therefore he must be 'Below Average', together with his militia commanders. On the other hand, Kalb, Williams and Gist, who commanded the Continental Line Regulars, displayed such courage and control that they merit the rating of 'Above Average'.

The British Regulars were first-class troops as were the American Continental Line Regulars. The American militia were ill-trained, hastily assembled groups of men whose morale and fighting ability was understandably low.

The British morale, together with that of the American Regulars, was first-class. The morale of the American militia should intially be second-class, rapidly dropping to third-class when they rout.

Although the terrain could possibly have had a considerable influence on the course of the battle, in the event it had no effect and the result would probably have been the same on any type of terrain.

Military Possibilities for this battle are almost non-existent, the prime possibility being that Gates should not have accepted battle in the first place! It is remotely possible that had he initially disposed his force with a stronger American left, with his Regulars forward rather than in reserve, it *might* have made a difference but this is highly debatable.

Guilford Courthouse—15 March 1781

This British victory marked the beginning of the end of the Revolution; its importance lay not in that fact nor in the numbers involved, the tactics employed nor in the casualties, but because by winning, Cornwallis so weakened his army that it caused him to lose the campaign in the Southern Colonies. Cornwallis, whose troops were perhaps the flower of the British forces in America—veterans commanded by experienced officers, had been pursuing Greene in the Carolinas in the hope of forcing an engagement whilst Greene, wishing to avoid battle, was endeavouring to draw the British as far as possible from their base. Cornwallis with 1,900 men and 3 guns advanced along the New Garden Road, crossing a small stream from which the ground rose gradually to a 500 yards square open space of three cultivated fields, bordered by rail fences, forming a defile between thick copses of trees. Greene, collecting reinforcements in Virginia that strengthened his force to about 5,000 men, had pushed westward to Guilford Courthouse and taken up battle stations, drawing up his force facing west in three lines 400 yards apart. With an excellent field of fire and protected by the rail fences, Greene had drawn up his first line of two brigades of North Carolina Militia, un-

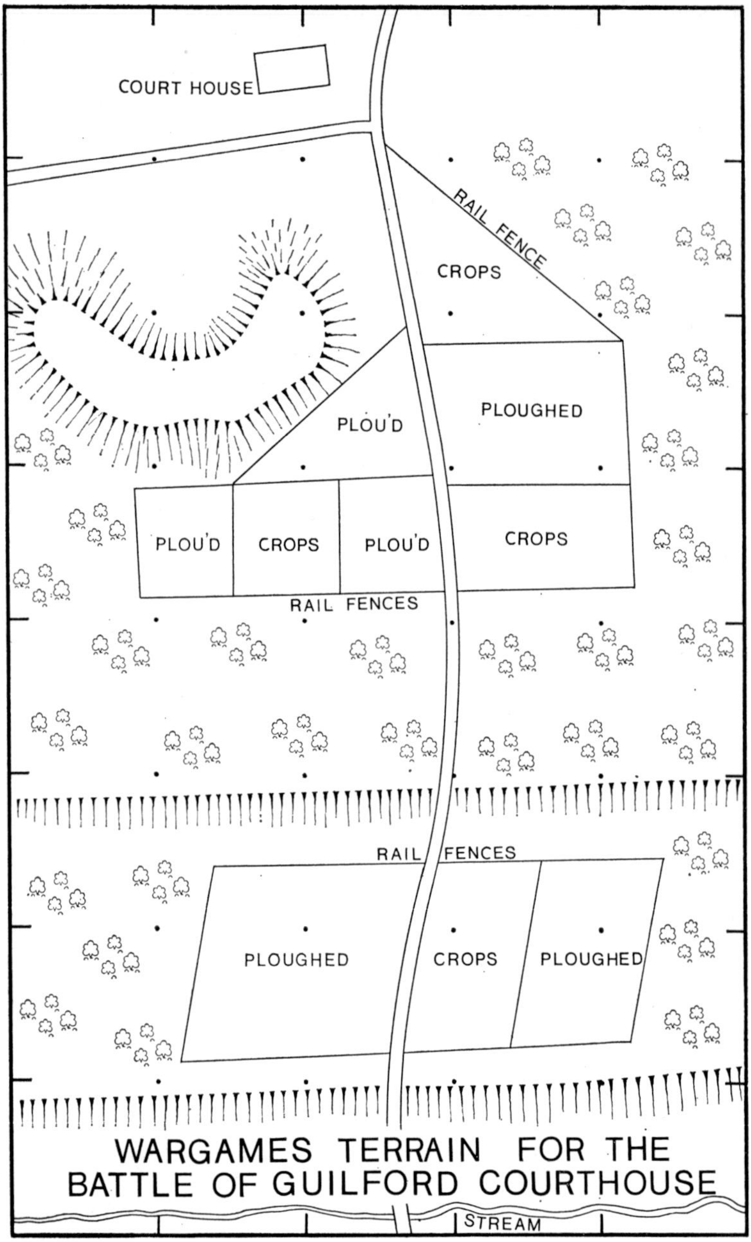

WARGAMES TERRAIN FOR THE BATTLE OF GUILFORD COURTHOUSE

trained soldiers without battle experience (1,600 men); on their left, he positioned two battle-experienced units—Lee's Legion (Regulars) and Campbell's Riflemen (frontiersmen from Virginia and the North Carolina mountains). On the right were Colonel William Washington's Regular Cavalry; some of the Delaware Regiment of Continentals and Lynch's Riflemen, veteran frontiersmen. Two 6-pdr guns in the centre on the road commanded the stream crossing.

About half a mile ahead of the clearing, in the woods that crossed in on the road, Greene deployed his second line of about 1,000 Virginia Militia, commanded by Generals Wilson on the right and Stevens on the left. Officered by men who had served in the Continental army and had some battle experience, this second line was somewhat stronger than the first. All were hidden in the woods on each side of the road, with picked marksmen behind them ready to shoot any man who ran away.

Then came another open space of cultivated ground, seamed by hollows and 400 yards in rear of the second line and rather to the west of the road was a hill that formed the salient angle in the midst of the clearing around the Courthouse. Near the eastern edge of this clearing Greene drew up his third line of 1,450 Regular troops, 2nd Virginia Brigade on the right and 2nd Maryland Brigade on the left, with two 6-pdr guns between them. The right flank was unprotected; the left flank, resting on the New Garden Road, was covered by artillery during the later stages of the battle. Both flanks of the first two lines were open, as the heavy woods only allowed frontal attack these exposed flanks were not a disadvantage.

On Cornwallis's right (under General Leslie) was Bose's Hessian Regiment and Fraser's Highlanders in the first line, with 1st Guards in support; the left wing was formed of the 23rd and 33rd Regiments under Colonel Webster in the first line, with the Grenadiers and 2nd Guards in support. A small corps of German Jagers and the Light Infantry were in the wood to the left of the road; Cornwallis posted his three guns on the road itself—they could not move anywhere else because the woods restricted their field of fire; Tarleton's cavalry remained on the road in the rear. After crossing the stream at the foot of the hill the British force formed up under American artillery fire losing a few men; the British artillery replied with an equally useless expenditure of ammunition. Amid a hail of fire from the invisible enemy in front and on the flanks the British troops advanced across the first clearing then charged with the bayonet, causing the North Carolina Militia to melt away in panic. Greene's two guns retired to take post near the road with the American third line. The British halted, their flank battalions turned outwards to oppose the American riflemen who were still pouring in a destructive fire from both sides.

Bose's regiment wheeled off to the right flank with the 1st Guards moving into the line to take their place next to the Highlanders; on the other wing the 33rd, with the Jagers and Light Infantry on their left, similarly wheeled off to the left flank; the Grenadiers and the 2nd Guards advanced to the left of the 23rd.

Slowly the redcoats pressed forward through the trees, driving the American riflemen back at the bayonet point; the cavalry of Lee and Washington fell back to conform. The regiment of Bose and the 1st Guards drove Lee and Campbell's left flank detachment backwards in a south-easterly direction, completely detaching them from the main course of the battle; this separate engagement was broken off by the Americans when the main conflict ended. Briefly, the American right flank detachment took position on the flank of the 2nd line, then retired to the flank of the 3rd line.

For about 400 yards, the attack proceeded east along the road and through the woods until it struck the 2nd line, where the rifle-armed Virginia Militia stubbornly held their ground and inflicted heavy losses to the advancing redcoats. Discipline and experience began to tell and gradually Lawson's Brigade on the American right crumbled before the Light Infantry and Jagers, who were their equals at bush fighting. Steven's Brigade, on Lawson's left, stood firm for a while but eventually both forces were pushed back until reaching the road, they broke and retired in disorder through the wood to Greene's third line, protected from pursuit by Washington's cavalry retiring with them to the edge of the forest.

North of the road, Greene's last line was opposed by the British left wing, who were slowed up by the heavy undergrowth and deep gullies. Led by Webster, the now considerably reduced 33rd, the Light Infantry and Jagers attacked the 1st Maryland Regiment. The finest battalion in the American Army, they steadily awaited the assault before pouring in a volley at close range, then charged with the bayonet to drive the attackers back in disorder, with heavy loss. Although severely wounded, Webster drew his men off into the shelter the woods and rallied them, supported by the three British 3-pdr guns of which advanced along the road and unlimbered on the riding ground at the edge of the forest, to keep up a steady and well directed fire.

While Webster's force were rallying, General O'Hara led the 2nd Guards, and the Grenadiers, supported by the 23rd and the Highlanders, against the Maryland Brigade. Seemingly giving way before O'Hara's attack, the 2nd Maryland Regiment steadily withdrew as the Guards pressed them; suddenly Washington's cavalry galloped out of the woods to crash into the rear of the Guards while the 1st Maryland Regiment flooded out of the undergrowth onto their left flank. Fighting fiercely, the Guards were utterly broken and the danger of a wholesale retreat forced Cornwallis to order his artillery to fire grapeshot into the struggling mass, causing great losses to both Americans and British.

In the pause that followed, Cornwallis reformed his line, rallied the 2nd Guards who had been joined by the 1st Battalion; Webster led up the 33rd with the Jagers and the Light Infantry on their left. Under the smoke of a volley, Tarleton's Cavalry went in to end the American resistance on the British right; then they moved across to join Bose's Regiment and press the Americans left flank, while the remainder of the British line engaged their front.

Seeing that the day was lost, Greene ordered retreat, abandoning his four guns; there was no pursuit by Cornwallis' exhausted troops. casualties were 93 British killed and 439 wounded; 78 Americans killed and 183 wounded.

Reconstructing the Battle as a Wargame

With the numerically superior American army in a position of their own choosing and the British attacking in their historically rigid formations, the only manner in which this battle can be accurately simulated is to devise 'local' rules that balance these factors. The Militia must check their morale state as soon as the British come within charge-move distance, or whenever they take fire and sustain casualties, or are brought into actual mêlée-contact. The British Light Infantry and the Jagers should be given the same powers of manoeuverability as the American Riflemen.

The cohesion of the rigid British formations should raise their morale otherwise it is doubtful whether they will ever get into mêlée contact, as they did in real-life. In fact, the morale of the British should be capable of being raised to great heights particularly in the case of the Guards, who were driven off after being severely mauled by the high grade Maryland Regulars, hit in the rear by cavalry and then flailed with grapeshot by their own artillery. Their subsequent rout was rallied and they actually came forward again to the attack—a most exceptional situation unlikely to be covered by the usual morale rules. It will be necessary to assess the varying states of morale and fighting ability of:

(a) British Regulars; (b) Continental Regulars; (c) American Riflemen; (d) American Militia.

The British Regulars and the Continental Regulars should be considered as top grade soldiers; the American riflemen used in the role of light infantry (but without the morale of the Regulars) while the Militia are low grade troops, both in morale and fighting ability, liable to flee the field without firing a shot.

Greene is certainly an 'Average' commander but could be classified as 'Above Average', although it will probably put victory completely beyond the reach of Cornwallis (himself an 'Average' commander) with his considerably smaller army. Webster could be 'Above Average' as could Washington; other commanders can be classified in relation to the performance of their units.

Military Possibilities can be devised to prevent the Militia from fleeing; to prevent the American second line from being pushed back; to alter the withdrawal of Webster's men and even their subsequent rallying; to allow the Guards to beat the Maryland Regiments and to turn and fight off the cavalry rear attack, etc., etc. But the historical reality of Greene's delaying action to maul his opponents, who consistently pushed forward indicates that the most realistic simulation will be achieved by judicious handling of the morale situation.

Construction of Terrain

Being fought in three separate stages, the required extensive table-top battlefield is provided by laying the terrain lengthways, with the stream on the 'bottom' baseline. From there an 18 inch wide ridge stretches right across the table, with three rail-fence enclosed fields each about 18 inches square. Fighting amid the trees requires a comparatively spacious area of woodland of irregular shaped hardboard bases to represent the entire wooded area, with occasional trees around the edges. The extensive ridge and hill are formed by draping cloth over positioned blocks or books.

Hobkirk's Hill—25 April 1781

Commanding the 900 strong garrison of Camden, Lord Rawdon decided to attack Greene's force after being told by a deserter of the American troops dispositions and the evacuation of their artillery. Subsequently Rawdon scraped together 800 combatants, including a group of convalescents, and prepared to attack Greene's position on Hobkirk's Hill. This sandy ridge running east and west amid the typical pine woods of the region had its west flank protected by a river and its east by the swampy bottom of a creek; towards the south the ground sloped 100 yards to a bushy plain and the hill itself was bisected by a road. It was not possible for an enemy approaching from the south to be seen from the hill until almost within assaulting distance.

Wargame Terrain for the Battle of HOBKIRK'S HILL, 25 Apr. 1781

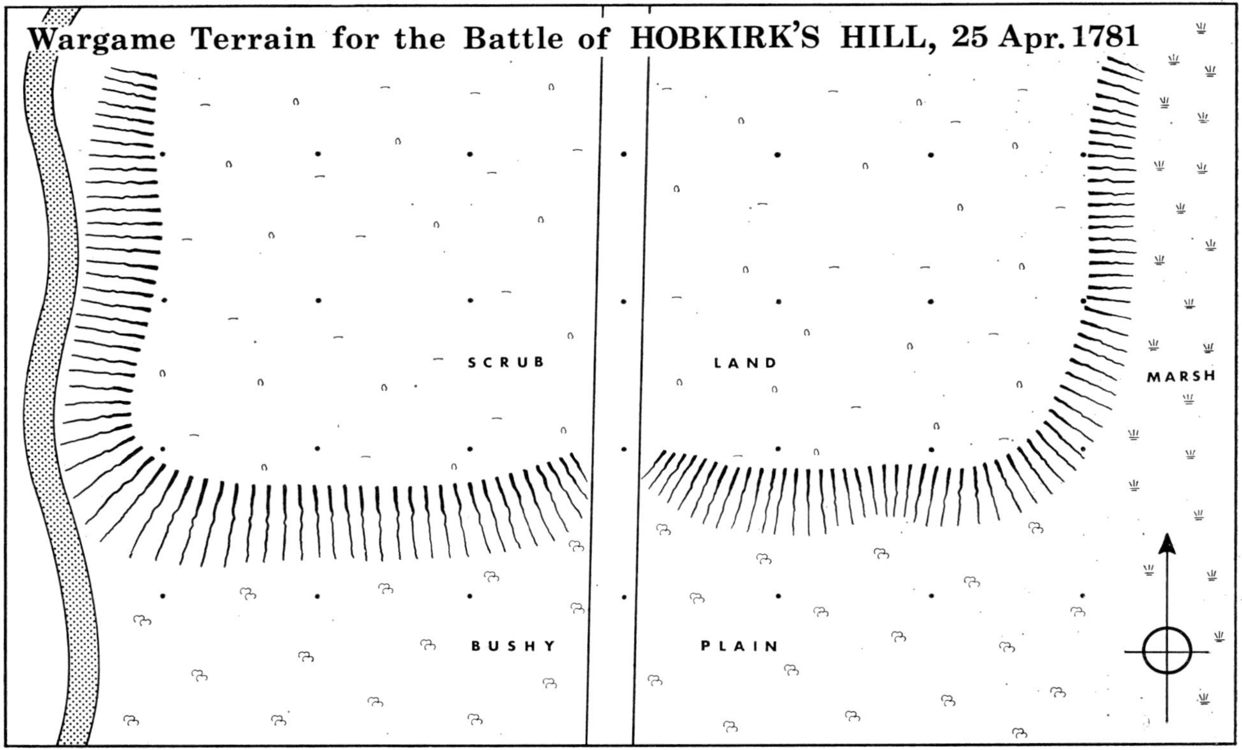

Numbering 1,174, Greene's forces were skilfully disposed to suit the terrain with the crack troops of Kirkwood's Delaware Company on the main approach from the south-east, fronted by two strong outposts commanded by Captains Benson and Morgan; other patrols covered the south and west approaches. On the west of the road, Huger's Brigade; Campbell's 4th Virginia and Hawes 5th Virginia Continental Line and to the east Williams Brigade, consisting of Gunby's 1st Maryland and Ford's 5th Maryland Continental Line. Washington's Dragoons (87 sabres) and Read's North Carolina Militia (250 men) were in reserve while 40 Virginia artillerymen arrived with three 6-pdrs in time for the battle and took up a concealed position on the road, to present an unpleasant surprise for Rawdon.

The British marched out of Camden and moved towards the relatively gradual south-east slopes of Hobkirk's Hill, then deployed for the assault with the front line consisting of the 63rd Regiment on the right, New York Volunteers in the centre and King's Americans on the left. Behind them on each flank were the Convalescents on the left, and the Volunteers of Ireland on the right, and to their rear the New York Dragoons and the South Carolina Tories. Seeing that they presented a relatively narrow front as they moved forward towards the Maryland brigade, Greene decided to attack Rawdon with a double development plan. His two centre regiments were ordered to attack frontally, while those on the flank enveloped the King's Americans and the 63rd Regiment; while Washington was sent to make a wide sweep down the road and hit the enemy rear. At first all went well with the British surprised and momentarily checked when the three American guns opened on them at short range with grape and cannister then, seeing Greene's advancing regiments, Rawdon moved his second line regiments up and extended the front line, so overlapping the Americans and threatening them with their own double development. Huger's Virginia Brigade gained ground against the British left but suddenly and without warning Gunby's veteran 1st Maryland Regiment, after halting to reorganise, fell apart in rout when charged by Rawdon's infantry. Almost at once the 5th Militia broke and the 4th Virginia followed them, leaving only Hawe's 5th Virginia unbroken to save Greene's army from annihilation. Holding fast they checked the British pursuit and only withdrew to avoid encirclement when the other regiments refused to be rallied.

The guns were saved after a gallant fight when a company of the Maryland Line sacrificed themselves to save the guns, before Washington's Dragoons rode back from their abortive action and hitched their horses to the guns.

Greene retreated two or three miles in good order while his rear guard checked pursuit and by late afternoon Rawdon's force had returned to Camden. By failing to destroy the American Army, Rawdon had gained nothing from his tactical victory.

Losses were about equal on both sides—266 American and 258 British.

Reconstruction of the Battle as a Wargame

Advancing from the expected direction and skilfully delayed by Kirkwood's crack Delaware Continental Line, Rawdon's dispositions were visible to Greene long enough for him to prepare a plan of action. Thus, this is virtually a 'set-piece' battle with both sides forces deployed facing each other. As soon as Greene moves forward to the attack, Rawdon should extend his own front line as he did in 1781. The unexpected presence of American artillery can be simulated by being hidden or not actually on the table (although marked on the map) until the moment of action. To conform to the general numerical scaling-down of forces, the three American guns should be reduced to a single piece—otherwise the odds, already heavily weighed against British, will be far too heavy to allow any chance of a realistic and authentic reconstruction.

In the American manner, Rawdon had placed Tory marksmen on his flanks to pick off American officers. On the table-top, this can be simulated by allowing each marksman to nominate his target-figure and then, a dice is thrown—if it falls 3 or 4 then the target-figure has been hit, if 1 or 2 then it is the man immediately on the target-figure's right; 5 or 6 it is the man immediately on the target-figure's left.

In a wargame between two markedly unbalanced forces, the smaller force will only have a fair and reasonable chance of success if certain 'local' rules and conditions are enforced. For example, the British commander Rawdon should be given the classification of 'Above Average' so as to raise the status of his force; the remainder of his subordinate commanders can be classified as 'Average'. Greene, the American commander and the ill-fated General Gunby will conversely be 'Below Average', with the remainder of their subordinate commanders 'Average'. The entire British force should be given an 'Elite' rating with appropriate bonuses in fighting ability and morale. With the exception of Hawes 5th Virginian Continental Line, who are 'Elite' troops, the entire American force will be classified as 'Average'.

The Elite status of the British troops will enhance their morale rating, while that of the Americans will initially be good. At the moment of truth when Gunby's 1st Maryland Line come up against the gleaming British bayonets their commander's low classification may reflect adversely (and realistically) on his regiment so, when subsequently deciding their morale state, they may rout as they did in real-life. 'Local' rules should provide for a chain-reaction among neighbouring regiments, probably with the sole exception of Hawes 5th Virginia, whose Elite classification should enable them to stand as they did at Hobkirk's Hill.

The terrain could have had a decisive effect upon the course of the battle had Greene, with his superior numbers, decided to adopt a defensive role. His decision to attack neutralised any effects the terrain may have had upon the course of the battle.

The obvious Military Possibilities arising from this engagement revolve around Greene's decision to attack—had he defended he would probably have won. And had Gunby's 1st Maryland Regiment held their ground and not routed, causing the remainder of the army to dissolve, again victory might well have gone to the Americans. Had these two factors favoured the Americans then another Military Possibility revolves around what might have happened had Rawdon attacked other than from the anticipated direction.

Cowpens—17 January 1781

For some weeks General Dan Morgan's small force had been pursued by Tarleton until, tired of running, the American commander decided to stand at Cowpens, an unconventional battlefield for which he planned an unorthodox deployment. Rested and fed, his men were positioned with a forward line of 150 picked militia riflemen—Georgians commanded by Major Cunningham and North Carolinians under Major McDowell—concealed in the grass and behind trees. These selected marksmen were ordered to allow the enemy to get within 50 paces, then to fire twice and move back to the second line, firing as they withdrew (to the astonishment of European soldiers, American riflemen could re-load and fire while running). Disposed along a crest for 150 yards behind them was General Pickens with the bulk of the militia (300 North and South Carolina troops) who were also told to fire and fall back after they were reinforced by the retreating riflemen from the front rank. They were ordered to hold their fire until the enemy were within killing distance and then pick off officers and sergeants before retreating round the left flank of the third line to re-assemble under its protection. Uphill 150 yards from the second line was Major Howard with the Delaware and Maryland Continentals in the centre, the Virginia Riflemen of Captain Triplett and Tate on their left and right, with Captain Beale's small company of Georgia Militia on the extreme right. Totalling about 450 men,

Wargame Terrain for the Battle of Cowpens, 17 Jan. 1781

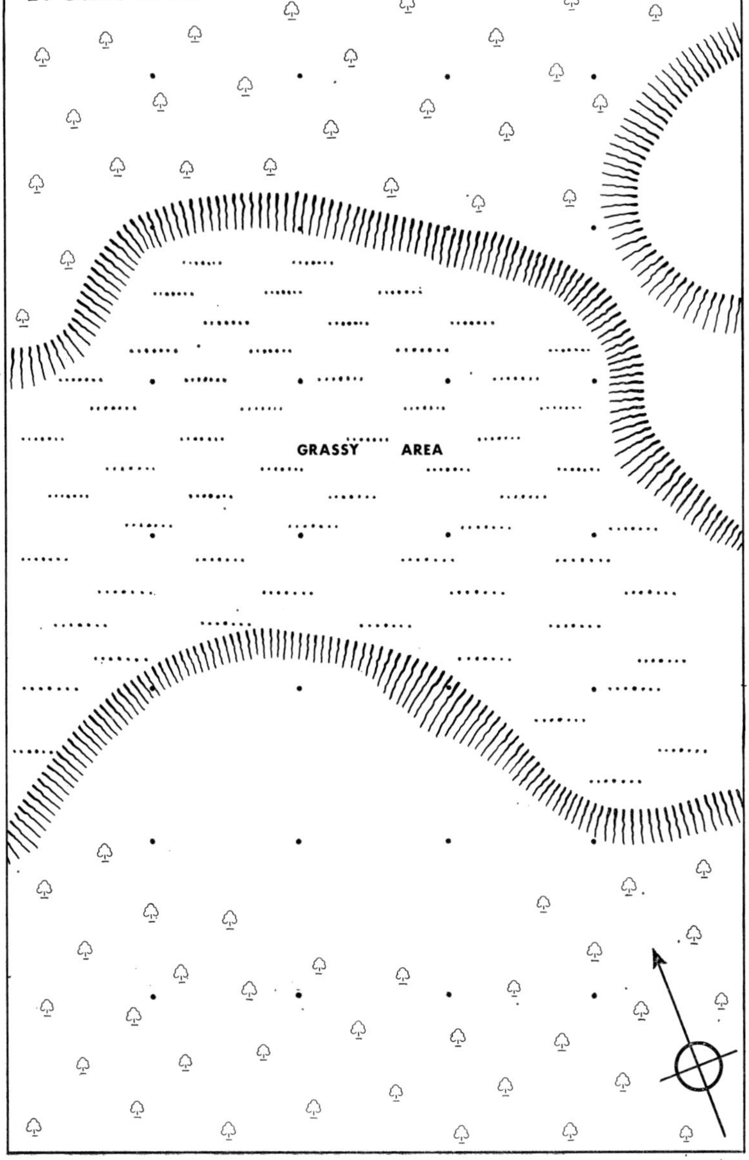

they held a front of approximately 400 yards. Washington's 80 Dragoons and Colonel McCall's 45 mounted Georgia infantry were posted as reserve half a mile behind the main battle position. The whole force totalled about 1,050 men.

Tarleton's force amounted to about 1,100 men, made up of 250 Infantry and 200 Dragoons of his British Legion; 50 men of the 17th Dragoons; 200 Highlanders of the 71st Regiment—all seasoned troops. There were 200 recruits of the 7th Regiment and a Royal Artillery detachment with two little horse-carried 3-pdr cannon known as 'grasshoppers', because they were mounted on legs instead of wheels so that the weapon jumped when fired. This particular pair of guns was to have an interesting history because they were captured from the British at Saratoga, recaptured at Camden, taken again by the Americans at Cowpens and recaptured by the British at Guilford Courthouse. Although Tarleton had a slightly numerical superiority, there was little to choose between the two sides because Morgan's militia were experienced marksmen and veterans of guerilla warfare.

After marching for four hours in darkness over difficult terrain, a patrol of British Dragoons came under fire, causing Tarleton to half his force and form them up for the attack some 400 yards in front of the first American position. Ogilvie's Dragoons spread out in a thin line and charged forward to drive in the skirmishers but the hidden marksmen sent them back with 15 empty saddles out of the 50 who charged.

Tarleton prepared to make a frontal assault, forming the 7th, the Legion Infantry and the three Line Infantry companies in a line. On each flank of the Legion a 'grasshopper' was ready for action to hasten the departure of the skirmishers who began to drop slowly back to the second line, firing as they moved. The 17th Light Dragoons were posted on the right flank of the line and another 50 Dragoons on the left with 200 Legion Cavalry in reserve and the 71st Highlanders 150 yards to the left rear of the 7th Regiment. With a Negro band setting the step, Tarleton personally led the line forward until they reached 100 yards from Picken's line. Then, reinforced by the rifle-men from the first line, the American Militia laid down a devastating volley causing numerous casualties that included 40% officers. Nevertheless the disciplined bayonet-tipped British line continued to come forward and the Militia began to run back to their left end of the third line as ordered. Seeing this the 17th Dragoons charged forward to cut them down but first they took accurate fire from Triplett's Virginia Riflemen then they were charged by Washington and McCall's cavalry. Outnumbered two to one the British Dragoons turned and fled, leaving 10 of their number on the ground.

Gathering themselves, Tarleton reformed his line of infantry and resumed the attack, the men going forward cheering until Howard's men, lying and kneeling for greater accuracy, poured volley after volley into them. This brought the British line to a halt but did not send them back and for the next thirty minutes a fire-fight took place between the two lines. Then Tarleton ordered the 71st Highlanders to envelop the American right and the kilted veterans started forward with pipes playing. Howard turned his right flank to meet this new threat, a move which caused the rest of the line to think a withdrawal had been ordered and the entire Continental Line moved back in good order to a new position. Seeing this the British line charged forward. The withdrawing American infantry turned, fired from the hip and then charged with the bayonet at the same time as the cavalry of Washington and McCall hit the British right flank and rear. The 7th Regiment recruits threw down their weapons and asked for quarter, while the veterans on the British right tried to make an orderly retreat but were rounded up by the American cavalry before they had covered 200 yards. On the left flank the Highlanders and Dragoons fiercely resisted until Pickens hit them in flank with some of his militia who had come over from the other side of the battlefield. Then, Colonel Jackson threw his Georgians into the fight and the surrounded Highlanders who, after a fierce hand-to-hand combat, were forced to surrender.

Tarleton galloped back to bring up his own Legion cavalry in an effort to save the day but all 200 troopers, ill-disciplined and spoiled with easy successes, rode off and left him. Throughout, the two small guns were hurling charge after charge into the enemy until the gunners were cut down almost to a man. With some difficulty, Tarleton himself made a fighting retreat and finally caught up with his 200 fleeing Dragoons and the party made their way back to the British camp.

In an hours fighting the British had 100 killed and 229 wounded with 600 captured; American losses were 12 killed and 60 wounded.

Cowpens was only the second occasion since Bunker Hill when it was seen what could be done by Militia who were good marksmen commanded by competent officers. Morgan's was a classic victory made possible through confidence in his men and their weapons which they handled with such devastating effect. It was a victory that raised American morale and attracted Militia recruits besides being a tactical reverse that led Cornwallis into fatal strategic errors at York Town.

Reconstruction of the Battle as a Wargame

Morgan fought this battle in a formal manner resembling the planned steps of a quadrille, based on his confident and accurate forecast as to the manner in which Tarleton would fight. If this battle is to be reconstructed with any degree of historical authenticity then both British and American moves must be rigidly programmed, with the historical dispositions and courses of action followed by both commanders reproduced in their order on the wargames table. It should make a good wargame because in a sense it bears a resemblance to the precise planning of many table-top generals!

Displaying outstanding personal leadership General Morgan must of course be classified as 'Above Average'. His subordinate commanders—Pickens, William Washington, and Howard were all similarly outstanding and must be classed as 'Above Average'. Tarleton and his subordinate commanders, whilst doing nothing foolish neither did they do anything brilliant—hence their classification of 'Average'.

Both the American and British Regulars were first-class troops and the militia, given confidence by their orders and friendly terrain, behaved very creditably. The cavalry of Tarleton's British Legion must, on their showing here, be classified as second-class both in morale and fighting ability. All other troops on the field were initially of the highest morale, with suitable fluctuations (mainly to the British) as the battle unreels.

Both sides fought in their characteristic manner which, in the past, had often led to American defeats. But on this occasion, Morgan skilfully utilised all the natural abilities of his men to bring out their highest fighting qualities on the day.

Selected by Morgan with the same sort of skill employed by the English commanders during the Hundred Years War when they sought ridges from which to best employ their incomparable archers, the terrain bore sufficient significant topographical features to be essential to Morgan's plan and tactics.

Employing such a progression of tactics with the success of each step depending upon what had occurred immediately before, the opportunities for Military Possibilities seem limitless. Had Morgan's plan gone wrong in any phase then he could well have lost the battle. Had the charge of the 71st Highlanders been successful against the American right (as British charges usually were in this war) then the American line could have been rolled-up. Even had Tarleton's cavalry obeyed orders and charged, it is doubtful whether, at that stage, they could have saved the day.

Eutaw Springs—8 September 1781

Taking a roundabout route, Greene surprised Colonel Stewart's encamped army at Eutaw Springs. Greene's cautiously advancing force, some 2,200 strong, consisted of Henderson's South Carolina State Infantry; Light Horse Harry Lee's Legion; Marion's partisans; Malmedy's North Carolina Militia and Pickens South Carolina Militia. Sumner's North Carolina Continental Line brigade; the Virginia Continentals of Colonel Campbell, and Colonel Williams' Maryland Continentals; Kirkwood's Delaware Continentals and William Washington's Dragoons. There were two 'grasshoppers' (3-pdr guns) and two 6-pdrs. The full breakdown of the American force was as follows:

Continental Line—1,256

South Carolina State Troops (Sumner's commanded by Henderson and Wade Hampton)—73 infantry and 72 cavalry.

North Carolina Militia—150

South Carolina Militia—307

Marion's force—40 cavalry; 200 Foot.

Lee and Washington—300 cavalry (200 men were detailed to guard the baggage).

Stewart's force of about 1,800 men consisted of the flank companies of the 3rd, 19th and 30th Regiments, forming a 300 strong flank battalion under Major Marjoribanks; eight battalion companies of the 3rd Foot; understrength 63rd and 64th Foot; Cruger's Tories and Coffin's South Carolina Horse. His artillery included two 6-pdrs, one 4-pdr, a 3-pdr and a swivel gun.

Warned by an outlying patrol of Coffin's South Carolina Horse, Stewart deployed forward of his camp in a line across the River Road that bisected the camp. Marjoribanks' flank battalion positioned themselves on the British right flank, in a dense blackjacket thicket that made them virtually invulnerable to the superior numbers of Greene's cavalry. Major Sheridan of Cruger's Tory corps occupied the Brick House with a detachment; Coffin's force of about 100 infantry and 50 cavalry were held in reserve.

Moving forward, the Americans brushed aside the British delaying force and then both sides came within musketry range of each other and a fierce fire-fight began. Both the American 'grasshopper' guns were put out of action and one of Stewart's guns was silenced. Greene's Militia showed up well, firing 17 volleys before showing signs of faltering, and in this phase of the battle Lee's Legion were fought off by the 63rd Foot when they tried to turn the British left.

Without actually being so ordered, Stewart's left wing moved forward and was followed by the rest of the line. Subsequently Malmedy's North Carolina Militia broke, quickly followed by the South Caro-

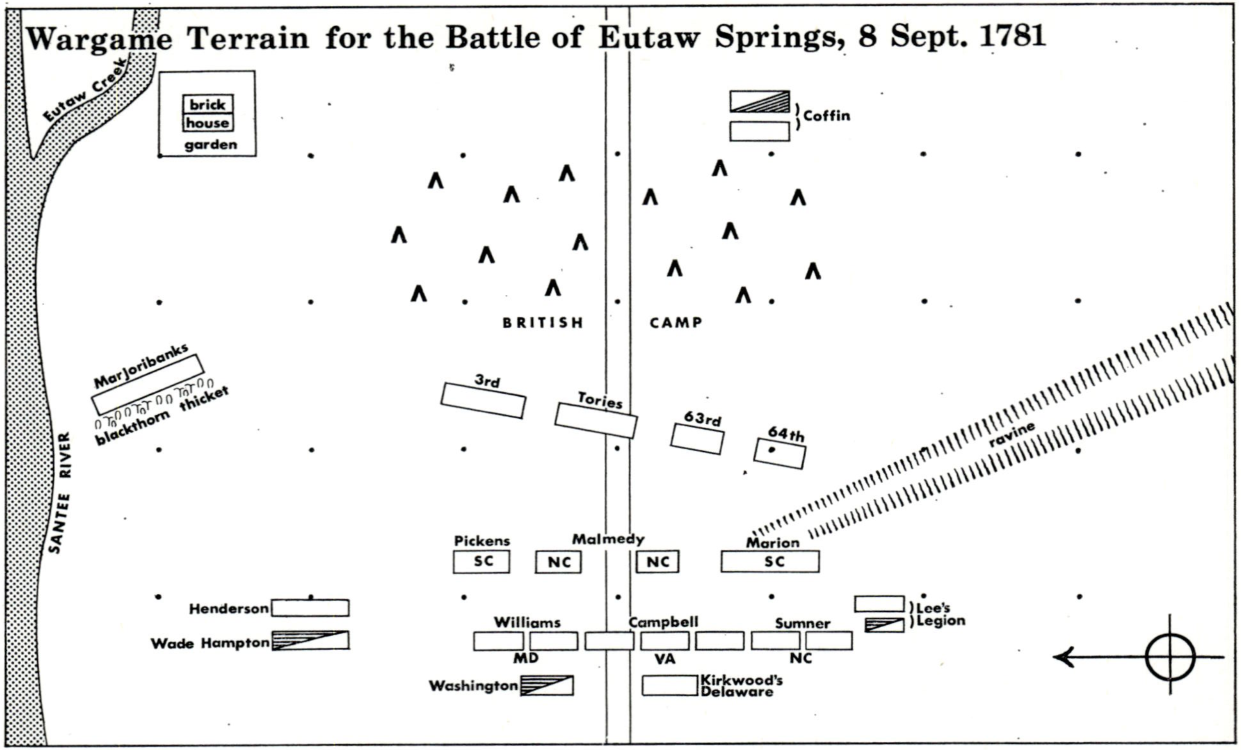
Wargame Terrain for the Battle of Eutaw Springs, 8 Sept. 1781
Eutaw Creek
brick
house
garden
Coffin
BRITISH
CAMP
Marjoribanks
blackthorn thicket
3rd
Tories
63rd
64th
ravine
SANTEE RIVER
Pickens
SC
Malmedy
NC
NC
Marion
SC
Henderson
Wade Hampton
Williams
Campbell
Sumner
MD
VA
NC
Lee's
Legion
Washington
Kirkwood's
Delaware

lina Militia on their right flank; Lee's Legion held their ground on the right against the 63rd, and the South Carolina troops on the other flank similarly withstood the British 3rd Regiment. At once, Greene brought forward Sumner's North Carolina Continentals to plug the gap and this force not only regained the lost ground but drove the British back to their original positions.

Stewart now brought up Coffin's cavalry to protect his left flank against the threat of Lee's horsemen, while Coffin's infantry reinforced the now shaken front line. Both commanders worked hard restoring their centres whilst fighting continued on both flanks. The American left, momentarily faltering, was rallied by Hampton's Cavalry who pushed back the 3rd Regiment and took 100 prisoners. Lee's Legion continued to hold the right but Sumner's North Carolina Continentals, green troops with only a month's training, gave way before an advance of the British centre. But Greene had not yet committed his best troops and now sent forward the Maryland and Virginia Continentals who delivered their first volley at 40 yards and followed up with a bayonet charge as the Legion infantry attacked the British left flank. Routed, the British left and centre retreated in confusion through their tented camp and the Buffs on the right, after holding on for a few minutes, fell back with them. It seemed as though Greene had won a hard-fought victory but Marjoribanks' flank battalion in their blackjack thicket held fast and wiped out half Washington's cavalry as they attempted to reach an open space to their rear. Wade Hampton charged with his South Carolina Horse and what was left of Washington's command, but they too were repulsed with heavy losses. Pursuing the fleeing British through their tents, Greene's men, both militia and Continentals were completely out of control as they gorged the food and drink that they looted from the British stores. Unaware of this, Greene personally conducted the left flank battle while the infantry of Lee's Legion held off Coffin's cavalry. But when he called for aid from the cavalry section of his Legion, Lee found that they had already been committed against Coffin from the American left and had been defeated. Next, Hampton's cavalry attacked Coffin and drove him back but in the subsequent pursuit the South Carolina horsemen were badly mauled by musketry as they passed close to Marjoribanks' position.

Now Lee's infantry moved against the Brick House, bringing two 6-pdr guns forward to batter Sheridan's citadel but both American gunners and infantry were driven back by musketry and a light swivel gun mounted in the house. The routed British troops now began to rally to reinforce Sheridan's strongpoint, some entering the house from the rear while others took up positions around the palisaded garden. Meanwhile, Marjoribanks' flank battalion had drawn back to the garden and had captured the rebel guns. Marjoribanks now led them forth in a counter-attack that turned defeat into victory, hitting the drunken rabble in the British tented camp and driving them into the woods. Marjoribanks was mortally wounded but other British troops reinforced his flank company and the battle was soon over with the Americans fleeing from the field on all sides.

The British suffered the highest percentage of casualties sustained by any force during the war, losing 693 men out of between 1,800 and 2,000 who initially took part in the battle. The Americans suffered over 500 casualties from the 2,200 engaged.

Eutaw Springs was the last major engagement of the war in the South and was one of the hardest fought battles of the entire Revolution. Both sides fought well with Greene scoring an initial strategic triumph and Stewart responding with an excellent deployment, particularly with his flank battalion placing, and the defence of the Brick House. In a day marked by great gallantry on both sides, the British simply out-fought the Americans in a confused action where ill-discipline snatched victory from Greene, although it was the fourth occasion on which Greene had won a campaign by failing to win a battle. Weakened by losses at Eutaw Springs, the British Army had to withdraw so that liberation of the South was virtually assured.

Reconstruction of the Battle as a Wargame

Greene's failure to completely surprise Stewart, allowing him to emerge from his camp and take up a position, really turns this into a 'set-piece' battle. Therefore, it should be laid-out on the table-top terrain with both forces deployed in their historical dispositions of 1781. When (and if) Greene's militia and Continental Line, driving the British back before them, reach the camp area then the conditions of the game must force them to halt for a specified number of game-moves, to represent a looting period.

Greene and his subordinate commanders can all be classified as 'Average'. For the British, Stewart and Marjoribanks both rate the classification of 'Above Average'—the former for his intelligent deployment and placing of Marjoribanks on his flank where this leader, at the cost of his life, saved and eventually won the battle for the British. The remainder of the British commanders are 'Average'.

The majority of Stewart's force were British Regulars and Cruger's Tories were veterans of the same calibre; only Coffin's cavalry were relatively inexperienced. For the Americans, the Continental Line were on a par with the British Regulars; the poorly trained and inexperienced militia held their own in fire-fights but could not face the threat of advancing British bayonets—this must be simulated by suitable morale rules when they are under actual attack. For the remainder of the game, after the enforced halt of Greene's troops while plundering the British camp, they must have their fighting ability penalised to simulate drunkenness and general disorder. To simulate the sterling feats of Marjoribanks' flank company they should be classified as Elite troops with appropriate bonuses in both fighting ability and morale-state.

The terrain played a major role in the course of this battle—lacking the presence of the blackthorn thicket and the Brick House the British would probably have lost. The river restricted out-flanking moves on the American left, while the ravine on their right affected their deployment and channelled their advance.

There are three potent Military Possibilities arising from this battle:

1. If Stewart had been completely surprised and lacked the time to leave his camp and deploy in a defensive position.

2. If Marjoribanks flanking company had not held on and then turned the battle with their counter-attack.

3. If the Americans had been better disciplined and not looted the British camp. In this they were not singular because, during Stone-wall Jackson's Valley Campaign of May 1862, the Federal General Banks was allowed to escape from Strasbourg because the crack Louisiana Zouaves and Ashby's cavalry stopped to loot a Union supply train; Napoleon's brilliant strategic penetration at the start of his first Italian Campaign nearly came to an abrupt halt when his troops were so busy looting the village of Dego that they were driven out.

ALL THE BEST GAMES HAVE RULES!

EVERYTHING THAT TAKES PLACE on the wargames table is controlled by rules designed to produce an authentic and realistic simulation of a battle. For this reason, the tactics and methods of the period must be known to the wargamer if he is not using commercial rules designed to cover these points. Well conceived rules bring extreme realism to table-top battles because they control the firing of different types of weapons and the resulting casualties, prisoners, fatigue, morale and other factors of warfare. The degree of realism is to the wargamer's choice; his rules can be very simple or very complicated. They come into effect as events begin to take place on the table-top battlefield, when the troops begin to move and the guns open fire. The conflict can begin in many different ways such as by drawing a diagram/map of the 'battlefield', to a scale of say 1cm on the map = 20 cms on the table-top. After studying the terrain and deciding their tactics, the opposing generals each mark on the map the opening dispositions of their troops. Then, having achieved a certain degree of concealment and surprise, the armies are set up and the battle begins.

A really hectic engagement follows when a curtain is drawn across the middle of the battlefield and, without being aware of the enemy's dispositions the opposing generals deploy their troops on either side. When the curtain is lifted there are usually some unpleasant surprises such as several squadrons of cavalry facing marching columns of infantry or a massed battery of guns!

Battles can be classified as *set-piece or encounter*—the former occurs when two armies marching towards each other find an area suitable to fight over, so they halt and form up in dispositions to suit their tactics. This type of battle frequently took place in the 16th and 17th centuries; Waterloo in 1815 was a Set-piece battle. An Encounter battle occurs when opposing forces come unexpectedly upon each other and begin fighting, as at Gettysburg in July 1863, a decisive

UNIT	OPENING DISPOSITIONS	1st MOVE	2nd MOVE	3rd MOVE	4th MOVE	5th MOVE	6th MOVE
Imperial guard	Centre - in village	Stay put F					
Chasseurs	In woods on right	↑ 12"					
10th Line	On hill on left	Stay put F					
Empress dragoons	Behind village	↑ 9"					
Cuirassiers	Behind left hill	↗ 12"					
Gun no. 1	In village	Stay Fire					
Gun no. 2	On left hill	Stay fire					

Fig 1 A Movement Sheet (filled in to denote the moves for each unit, as the game progresses)
1 inch = 2·54 cm

battle of the American Civil War. It is possible to simulate almost any type of battle on the wargames table, such as small numbers of men holding a defile or a bridge against superior numbers of the enemy; attacks on supply columns, cavalry raids, sieges, amphibious operations, etc., etc.

Certain basic factors are common to all rules—cavalry move faster than infantry and both move more slowly when climbing hills or across rough country. Men sheltering behind stone walls or in houses, because they are not as vulnerable to fire as when out in the open, should be given a bonus. Rules must indicate how long (usually expressed in distance—i.e. a part of the move is forfeited) it takes to cross deep and fast flowing rivers when units are channelled onto bridges, or the shallow streams that are fordable. Similarly, an agreed distance is deducted from the move when walls and hedges intervene.

The controlled movement of armies takes place in a variety of ways. The winner of a dice-throw decides whether to move *before* or *after* his opponent, bearing in mind that the army that moves second fires first, before their opponent. So it has to be decided whether it is of greater advantage to move one's troops first and to receive the enemy's fire before you can return it. Then there is the *Simultaneous Move* method that allows one army to move their units beginning the right-hand side of the table while the other army starts from the left-hand side, with each general deploying his troops while anxiously watching the activities of his opponent at the other end of the wargames table! Simultaneous moving requires, before the move commences, for each commander to write his intentions on a ruled form (See Fig. 1) with an arrow pointing in the direction the troops are to move, or he writes '*Fire*' to indicate the firing of muskets or artillery. If actual contact is intended then '*Charge*' is written with an arrow indicating the direction of the charge and the unit that is being attacked.

Best controlled by an umpire, another method of moving consists of specific orders for each unit being written *before the battle commences*—thus the 2nd Maryland Regiment are ordered to '. . . march down the road, turn right at the crossroads and attempt to capture the hill'; or 'Light Horse Harry' Lee's cavalry '. . . will move out on the right flank and engage enemy cavalry opposing them'. Units must conform to these orders, not deviating unless the orders are countermanded by the commanding general. Difficult to handle and requiring honesty from both commanders, this is perhaps the most realistic method of controlling wargames.

The move-distances allowed wargames figures plays an important part in the development of the game—short move-distances result in a slow and halting battle while over-long move-distances makes the game a wild and sprawling affair. So a balance has to be struck to ensure a reasonably fast-flowing and realistic game. On a wargames table of 250 cm × 150 cm in size, the following move-distances will be found to be suitable.

Infantry —25 cm per move.
Cavalry —40 cm per move.
Guns (horse-drawn)—40 cms per move, allowing 10 cms for unlimbering or limbering up.

To represent the firing of muskets or rifles, one volley of rifle fire is permitted for every five men for whom a single dice is thrown. Thus, a regiment of 20 men will fire four volleys by throwing four dice; if the regiment is down to 15 men then three volleys are fired. Assuming a reasonable maximum range for musketry to be 60 cms, the following variations allow for differing ranges:

At 60 cms (long range) —Deduct 3 from each dice.
At 30 cms (medium range)—Deduct 2 from each dice.
At 15 cms (close range) —Deduct 1 from each dice.

Deduct 1 more per dice if the target/men are sheltering behind a wall or in a house.

In practice, say 20 men are firing at 40 cms (long) range so that 3 is deducted from each dice; the first dice scores 6, resulting in 3 enemy casualties; the second dice scores 3 and so causes no enemy casualties; and third dice scores 4 which means 1 casualty and the fourth dice scores 5 or 2 casualties—making a total of 6 enemy casualties. If the enemy are under cover, such as in a house or behind a wall, another 1 has to be deducted from each dice score—allowing only 3 casualties in this case.

When firing on cavalry it is possibly more realistic to only count half-casualties, as some of the bullets may hit the horses who are strong enough to carry on when sustaining a wound that would bring down their rider.

From its earliest days up to the middle of the 19th century artillery guns were unrifled muzzle loaders, far from accurate although often most expertly handled by experienced gunners. To allow for such imperfections, and with a maximum range of say 100 cms, a dice is thrown each time the gun is fired. At ranges between 50/100 cms, 6 must be thrown to hit; from 25/50 cms—5 or 6; at ranges below 25 cms—4, 5 or 6. When a hit is scored, throw another dice to indicate the number of casualties. Half the number of casualties can be counted when firing on cavalry or men behind a wall or in a house.

A realistic method of simulating the firing of explosive shells is to throw a dice to secure a hit and then to place a 'burst-circle' with its centre at the point of the shell-burst. A 'burst-circle' is a wire ring or a transparent plastic circle 5 cms in diameter—every man covered by this circle is considered to have been killed, or the circle can be 8 cms in diameter with a 4 cm circle marked in its centre. Everyone within the small circle is killed whilst those within the outer circle are considered wounded and out of action for say two moves. Or use a 'group' of numbered circles, the casualties are those men in the numbered circle indicated by a dice throw (see Fig. 2). Very close range artillery fire (grapeshot) is simulated by using a wire 'funnel' shape, extending 30 cm from the gun barrel as in the illustration (Fig. 3). Everyone within its wire borders is considered killed.

When two forces contact each other in hand-to-hand fighting it is called 'a mêlée'. A straight forward affair between groups of infantry or of cavalry can be settled by throwing one dice for each group of five men with an agreed proportion of the total score representing the number of enemy casualties. Thus, 20 Red men are fighting 15 Blue men—the Red side throws four dice and scores 4, 2, 1, 5 which, at 'full rate' equals 12 enemy dead. The Blue force throw three dice and scores 6,4,5 making a total of 15 enemy casualties. If it is agreed that *HALF* the dice-score should count, then only half the number of the casualties would have occurred. It is usual for the side sustaining the most casualties to be forced to retreat an agreed distance from the area of the melee. In 'Mixed' melees between cavalry and infantry, the cavalry are considered to have a 'shock' effect because of the force of their charge, represented by adding 1 to each of the cavalry dice. In these mixed mêlées, cavalry count 2 points each and infantry 1 point; in a mêlée between 10 cavalry and 10 infantry the cavalry (20 points) would throw four dice in all. The infantry (10 points) would only throw two dice (one for every 5 points). At 'full' rate, if the four cavalry dice score 2, 2, 1, 3, they have killed 8 enemy infantrymen, while the two infantry dice score 5 and 3 making a total of 8 points in cavalry casualties, or 4 men. So the cavalry win the melee and the infantry retire—to be chased by the cavalry in the following move!

A very simple and basic method of deciding a mêlée is to rule that the side with the most casualties retreats, but a more realistic method is to consider the MORALE of the opposing forces—Napoleon considered morale factors to be three times as important as physical! Adequate morale rules that produce reasonable and realistic effects upon wargames regiments are not difficult to simulate. The circumstances and conditions which frightened or encouraged men have not changed since the beginnings of Time, so there are certain basic morale factors common to all periods of history. Morale giverns the discipline and confidence of soldiers, so that wargames rules must allow troops to manoeuvre in accordance with the state of their morale—whether they are high-class troops showing bravery and staying-power, or low quality soldiers who break and flee at their first casualties. As an example, at the end of our demonstration mêlée there were two infantrymen confronted by six huge horsemen. Whether they bravely stand or flee can be decided by each General throwing a dice and multiplying his men remaining after the mêlée by the score on the dice. So, if the cavalry General scores only 1 for his six survivors then their morale figures is 6; should the two infantry score 4, 5 or 6 then they have won! This possibly unrealistic victory against greatly superior odds reflects their high morale, aided by luck—very evident in warfare!

The amateur General, armed with a knowledge of the most elementary aspects of movement, fire and hand-to-hand fighting can fight a battle that may not shake the destiny of nations but will be part of a stimulating and pleasurable table-top military career. All 'amateurs' have dreams of becoming 'professionals' and wargaming is no exception so there will soon be a craving for more realistic battles, requiring control by rules that, in the case of the American Revolution, simulate such authentic factors as the differing tactical and morale standards of Militiamen and Regulars, while allowing the American commander to employ characteristically flexible tactics as

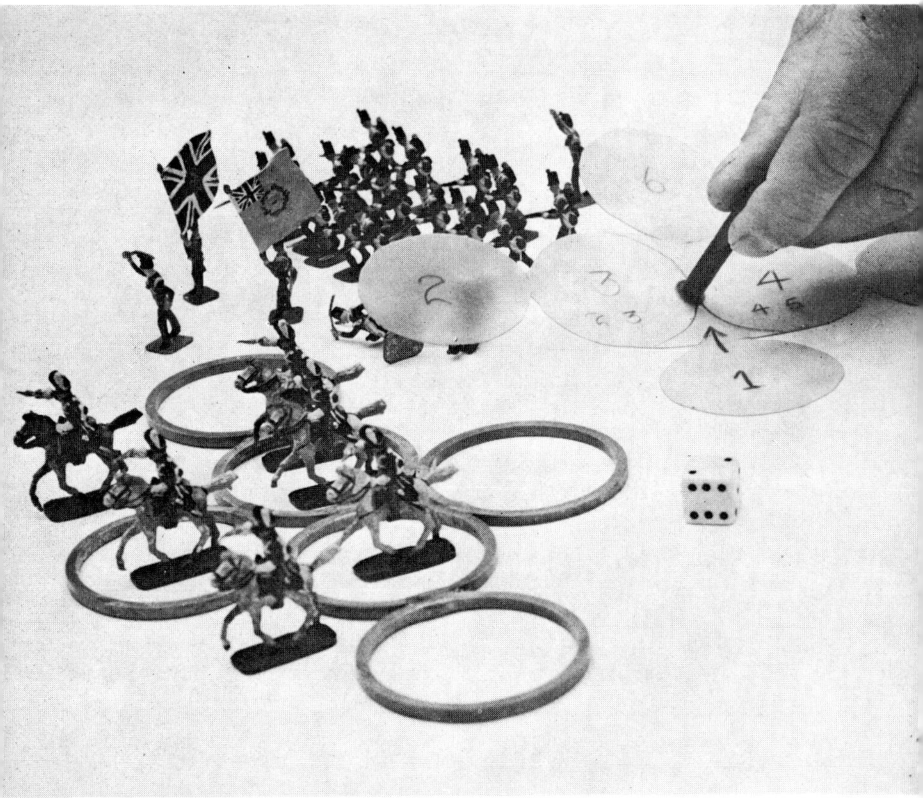

Fig. 2 Method of simulating artillery fire—see page 79.

Fig. 3 Method of simulating artillery firing grapeshot—see page 79.

Fig. 4 Content of the Airfix set S 39/59 *Washington's Army* (Photo provided by courtesy of Airfix Products Ltd.)—see page 91.

Fig. 5 Contents of Airfix set S 40/69 *The British Grenadiers* (Photo provided by courtesy of Airfix Products Ltd.) see page 91

Fig. 6 Lifesize figures of American Revolution combatants (from the *Sunday Times* '1776' Exhibition).

Fig. 7 British infantry advance during a wargame in the American Revolution period.

American Line Infantryman in hunting shirt.

American Continental Line Infantry.

British Light Infantry.

American Rifleman in hunting shirt.

Baylor's Dragoon's, Officer.

American Frontiersman in coonskin suit.

Washington Guard, Officer.

New Jersey Regiment, Drummer.

Continental Line Infantry, Drummer.

Green Mountain Ranger, Officer with spontoon.

Green Mountain Ranger, Officer.

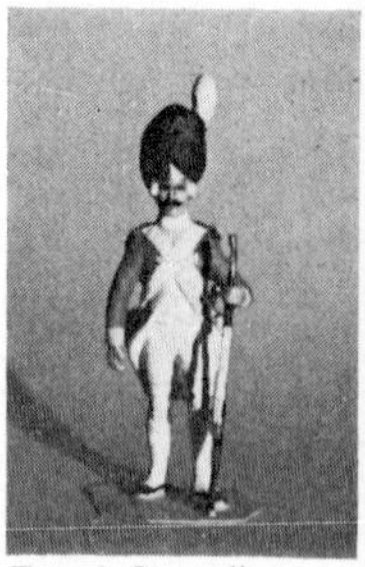

French Grenadier.

French Musketeer.

British Grenadier Officer, 5th Foot.

British Grenadier, Officer, 57th Foot.

Hessian Grenadier.

Fig. 8 Model soldiers of War of Independence regiments.

Hessian Musketeer.

Hessian Fusilier.

Hessian Artilleryman.

Pennsylvania Rifleman.

Margonis Rifleman.

Massachusetts Militia.

3rd North Carolina Regiment.

Officer, Continental Line Infantry.

Maryland Line Infantry, Officer.

British Grenadier, private.

American Loyalist Light Infantry.

Lafayette Light Infantry, Officer.

Pennsylvania State Marine.

King's Royal Regiment of New York

6th Continental Line Infantry.

2nd New Hampshire Regt.

Fig. 8 Model soldiers of War of Independence regiments.

Fig. 9 The British Regiment's colour-party take fire from Lafayettes Light Infantry, resting their muskets on the rail-fence—scene from a typical wargame.

Fig. 10 British grenadiers surprise some American skirmishers among the ruined farm buildings—scene from a typical wargame.

his British counterpart is restricted to the rigid manoeuvres of the period. Accounts of battles of the American Revolution reveal the varying, often poor quality of the American Militia units. When reconstructing table-top simulations of 1776-1781, allocate a 'Panic' rating to each unit:

Militia	—5 points
German Mercenaries, American Infantry	—6 points
British Regular Infantry, Cavalry	—7 points

To cause a unit to panic, this points score has to be *doubled* by throwing a dice and adding the score to the Panic-triggering figure. To that figure, add ONE if the unit is facing a bayonet charge and TWO for a cavalry sabre charge. So, if a Militia unit facing a British bayonet charge (which adds 1) throws a 4, 5 or 6 then it reaches its double-figure of 10 and makes off, in disorder, for a prescribed distance.

Here are a set of rules that allow a realistic and fast-moving wargame in the period of the American Revolution.

Basic Rules for Conducting Wargames in the American Revolution Period

These rules are compiled to cope with infantry regiments of 4 officers and 40 men; cavalry units of 2 officers and 18 men (or squadrons of 1 officer and 10 men).

Moves

Infantry (with exception of those in next section)	12″	Without firing; can move 6″ and fire once or 2 volleys if not moving.
(Count $\frac{1}{2}$ value for firing or mêlées when moving in woods)		
Light Infantry	15″	May fire 1 shot when moving full distance
Riflemen, Indians		May fire and then move in move-fire-move
Cavalry (cannot move in woods)	18″	
Guns (cannot move in woods)	12″	Move 12″ without firing move 6″ and fire 1 round No move and fire 2 rounds
Generals and Staff Officers	15″	

Deduct 3″ from move for:

Limbering guns
Unlimbering guns
Mounting or dismounting cavalry
Forming column into line or vice versa (or generally changing formation)
Crossing a wall, hedge, or fording a river
Moving in woods (NOT light infantry, Riflemen or Indians)

Move $\frac{1}{2}$ speed up hill
Add 3″ when moving on recognised roads.

With the exception of Light Infantry; Riflemen and Indians all infantry units move and fight in close order.

Morale

When a unit has been in action (or taken fire) they have to test their morale to see if they are 'shaken'. Throw 3 dice and make the following additions and subtractions:

(a) 2 is added when Commanding Officer is with the unit.
(b) 1 is added for each officer or standard-bearer present.
(c) 3 is deducted if a unit is below half-strength.

Militia units *ALWAYS* deduct 1 from *ALL* dice.

A unit DEDUCTS 1 from each dice score on the *first* occasion it has casualties; Militia units deduct 2. To avoid being shaken, a unit must total 12 with its 3 dice.

The presence of a Staff Officer with a unit will add ONE to the dice score. C-in-C adds TWO.

A unit that is retreating in rout has to be rallied in its next move before it can resume. A dice is thrown before it moves:

(a) At first time of rallying—4, 5 or 6 will make them turn.
(b) At second time of rallying—5 or 6 will make them turn.

Otherwise they continue to retreat 12″ in the same direction.
If fired upon or pursued—a unit requires (a) 5 or 6 (b) 6.

When friendly units on either side (or in front) are breaking and fleeing, then a unit has to find out the state of its own morale, by scoring *12* with 3 dice, if successful, then it stands.

When a unit fails to makes its '12' a dice is thrown:

5 or 6—the unit rallies.
3 or 4—the unit retreats 12″ ends up facing enemy.
1 or 2—the unit retreats 12″ ends up backs to enemy.

In the latter case, the unit has to be rallied before making any further moves.

Elite units ALWAYS ADD one to each dice at their initial throw.
Overwhelming Numbers: When a unit is faced with numerical odds of two to one or more then it throws a dice:

5 or 6—carry on.
4 —half unit fights; half runs and need saving-throws.
3 —unit runs 12″ ends up facing enemy.
2 —unit runs 12″—throw for each man. Survivors end with backs to enemy.
1 —unit cut down without saving throws.

When C-in-C is killed—immediately dice for all units within 24″.

5 or 6—unaffected.
4 —fall back half full-move distance.
3 —fall back full move.
2 —unless rallied by Staff Officer next move will march off field.
1 —unit lays down its arms.

When standard-bearer is killed in mêlée, his colour will be captured if enemy throw 4, 5 or 6.

Artillery Firing

Crews: Guns have 4 men crews. They may be re-crewed by infantry in one complete move.

If two men of a crew are killed IN THE SAME MOVE, that gun may lay OR fire but NOT both in that particular move.

A gun with only 1 man remaining may only move *OR* lay *OR* fire.

Guns are laid *after* both sides have completed their moves.

Guns can fire 2 rounds of round-shot or grapeshot in a move.

Guns may only fire upon troops going into a melee when those same troops have crossed the gun's *original* line of fire.

Guns have a flat trajectory and may not fire over the heads of troops, unless sited on ground higher than those troops.

Guns sited on a recognised hill add 6″ to their range.

Guns firing upwards on targets situated on a hill or on walls, etc., deduct 6″ from their range distance.

Captured guns may be used against their former owners—it takes one complete move to thus re-crew a gun and it may not fire during that move.

Ranges of Guns: 36″ and 18″ grapeshot.
Method of firing:
Nominate exact point of aim.
Two dice are thrown (one white, the other coloured).

Score on *white* dice: 1 = shot falls short of target.
2 = falls short over target.
3 = falls short to right of target.
4 = falls short to left of target.
5 = Hit on target.
6 = Hit on target.

The score on the coloured dice = number of inches shot falls from target.

For casualties by grapeshot (18″ range). Lay a 'grapeshot shape'—triangular wire shape 18″ × 2″—from the gun barrel—everything in the front half of the shape can be saved by normal saving-throws. Every man in the rear half of the shape is killed without saving-throws.

Gun firing on gun—A DICE IS THROWN TO SEE if crew or gun are hit—1, 2, 3 it is the crew and 4, 5, 6 it is the gun. If it is the gun then it is considered that a wheel has been smashed and it will take 1 complete move to repair—during which time the gun cannot move or fire.

Gun fire on walls and houses—Stone or brick houses cannot be destroyed by gunfire. Wooden houses are destroyed after 2 hits, but provide ½-cover (i.e. only ½ casualties to occupants). Men in stone houses or behind walls do not take casualties from gun fire providing they do not expose themselves to fire their own weapons. Militia units ALWAYS deduct 1 from all firing-dice scores.

Infantry Fire

Two ranks fire. As far as is reasonably possible, units will only fire upon, and be fired upon, by their opposite numbers.

Method: 1 dice represents a volley of five men firing; odd numbers are catered for by 3 or more men counting as a volley.

Men firing from houses or from behind hard cover add ONE to each dice. They only suffer HALF casualties when fired upon.

Range	Troops
12″	¼ total dice score
8″	½ total dice score
4″	¾ total dice score

Light troops, riflemen and Indians fire individually—throwing 1 dice per man, adding 1 to score if firing from cover.

18″	6
12″	5 or 6
6″	4, 5 or 6

Indians have 'trade' muskets and each man throws dice for a misfire indicated by 1, 2 or 3.

Saving Throws

At the end of a move, throw a dice for each casualty. Scores on chart below indicate that man is only lightly wounded and can carry on. Lower scores mean that he is killed and removed from battle.

Arm	Formations in the open	Lt. Inf; Riflemen; Indians	Hard Cover Houses	Hedges
Infantry	5, 6	4, 5, 6	4, 5, 6*	4, 5, 6 (musketry only)
Cavalry	5, 6			
Gun Crews	5, 6 (side)	4, 5, 6 (front)	3, 4, 5, 6	

Order of firing: This can be simultaneous i.e. all weapons are considered to be firing at one and the same time thus avoiding any advantages through priority of firing.

Or else it can be done on a 'chance' basis by throwing a dice for each and every unit or weapon. In this way, unit 'A' facing unit 'B' will throw a '6' whilst unit 'B' only throw a '4'—so unit 'A' fires first then the survivors of unit 'B' have their firing. Militia units deduct 1 from their dice-score.

An improved method is to make sets of small squares of card bearing numbers from 1 to 6—about twenty of each number. After the move has taken place, draw a handful of these cards and without looking at the numbers, place one by the side of each unit or gun. When all cards have been laid, begin with the 6's and fire them off, then the 5's and so on.

Light Infantry (including Riflemen and Indians)

May split-move-thus they can move part of their distance—fire—then move remainder. Enemy target may fire upon them in return if they throw 4, 5 or 6.

Light Infantry and riflemen in open order are considered to be a skirmish line and may be fired through by their own troops and guns.

Light Infantry, riflemen and Indians fire individually and may select individual targets, stating how many of them are firing upon it. Thus, if five men nominate an officer, and the first man kills him, then the other 4 have wasted their shots.

Such hits on nominated targets have to be diced for—thus, a dice is thrown:

3 or 4 means that target man is hit.

1 or 2 means man nearest his right hand.

5 or 6 means man nearest his left hand.

The usual saving-throws prevail in these cases.

Light troops, riflemen and Indians who are attacked may refuse mêlée if they have sufficient move-distance in hand (i.e. if they have moved only a proportion of their possible move-distance) to get out of reach of attackers. If they are not able to do so, then mêlée takes place in usual way.

Mêlées

Method: A mêlée lasts for two rounds per game-move.

Front ranks only fight the first round of a mêlée and are reinforced by the second rank during second round of mêlée.

Mêlées are between the men themselves, thus each man will individually fight his opposite number (if more than 1 then he dices to see if he can dodge 2–1 needs 4, 5 or 6 to do so). Throw dice as under:

American Militiaman British/American Artilleryman	—1 dice	
American 'Continental' Infantryman Riflemen Indians British/American Light Infantry	—$1\frac{1}{2}$ dice	(3 dice to every 2 men)
British/German Regular Infantry Cavalry	—2 dice	

When an attack is actually led from the front by a Staff Officer no mêlée-dice fall less than 3. If C-in-C leads then no dice fall lower than 4.

Infantry behind hard cover add $\frac{1}{2}$ dice score as long as attackers are on opposite side of such cover.

Casualties are taken from the front rank; men exactly opposite each other cannot both be killed so that removing a man leaves a gap. Into this gap the triumphant attacker will move so that he takes up the position originally occupied by the man he has killed. The defender who kills his opposite number has the option of doing the same. This ruling does not apply when defenders are behind a wall or in a house.

Casualties are 'saved' by throwing a dice for each one; 5 or 6 saves an infantryman; 4, 5 or 6 saves a cavalryman.

Staff Officers and Generals may be considered casualties when the opposing side throws a 5 or 6 in the mêlée-scoring throw; either one may count as personally against a Staff Officer rather than kill line troops. The Staff Officer is saved then by 4, 5 or 6.

At conclusion of Mêlée: Count infantry/artillery as 1 point; cavalry as 2 points. The force suffering the most casualties (in points) after two rounds of fighting throw to try to make their '12'. (see 'Morale' section). This is also done by any unit that has no officers left standing.

The presence of a Staff Officer actually with a unit adds ONE to the first dice throw. C-in-C adds TWO.

If they do not make their '12' then they are considered to be shaken and throw a dice again—acting in accordance with the following table:

5 or 6—they rally and carry on fighting next move.
4 —they retreat 12″ and are pursued by enemy—mêlée continues next move.
3 —they retreat 12″ in orderly fashion and end up facing enemy.
1 or 2—they retreat 12″ in rout and end with backs to enemy, who pursue and carry on fighting with benefit of a rear attack.

Units falling back after a mêlée cannot move in the next game-move; they had their move when they fell back. They may face or turn if attacked (will need rallying first if in rout) and will fire and fight if attacked.

Victorious Side after Mêlée: When defeated side has retreated winners may move 3″ to tidy up their ranks. If retreating side are routed, then victorious side throw a dice—1, 2 or 3 they are out of control and dash forward a further 6″ after the routed enemy. If this means a second mêlée in the same round then it goes on. If a force has a General with them then they do not get out of control.

The defeated force falls back in strict accordance with their final position in the mêlée—this means that each individual's move back is measured separately.

At the following move, the winner of a mêlée will deduct 3″ from his move-distance. This is to allow for him reforming. He may move without doing this but each man's forward move has to be measured individually so that they move in disorder.

A unit completely surrounded at end of mêlée will need to total 15 instead of the usual 12.

Cavalry Attacking Infantry

1. When the cavalry take casualties during their approach they try to make '12' with 3 dice and, if unsuccessful, they will not charge home

but will halt at the point where they last took fire—allow 3″ for reining-up—then retire the remaining portion of their move. They will end up facing the enemy.

If the cavalry charge-home then the infantry throw to see if they withstand the charge—they will try to make their '12'. 1 is added to dice for every rank over two. Infantry will deduct 1 per dice if in open order or in single rank.

2. When cavalry charge home on infantry from a frontal position throw 1 dice for each cavalryman in front rank of charge; cavalry will kill a man for all throws of 1, 2, 3 or 4.

The infantryman so removed will leave a gap into which the cavalryman is pushed forward so that he takes on the man in the second rank. If there is no second rank, the cavalryman moves on, takes 3″ to turn and then takes part in mêlée from the rear.

3. If an infantry unit does NOT stand—a dice is thrown for every infantryman and he runs rearwards 2″ for every 1 on the dice. Pursuing cavalry automatically kill all men they can reach within the limits of their move-distance.

Attacking Artillery: If a gun is overrun by attacking infantry or cavalry a dice will be thrown for each gunner:

1, 2, 3—cut down
4, 5—flee
6—spike gun and retire

Attackers may spike a gun by remaining with it for a complete move. They can remove gun if horses are available.

Infantry Attacking Infantry

1. When an attacking unit takes casualties during their approach they will throw three dice and try to make 12. If unsuccessful they will turn at the point where fire was last received and move rearwards for the rest of their move-distance. They end up facing the enemy.

2. If attackers make their '12' then the defenders need to make their '12' to stand.

3. Should defenders NOT stand, dice again:
1 or 2—defenders routed 12″ with backs to enemy.
3 or 4—defenders retire 12″ facing enemy.
5 or 6—defenders choose between moving back 6″ or 12″ (facing enemy in each case).

These distances count as their NEXT game-move.

4. All defenders add ONE to their dice when making their '12' if they are behind cover (1 to each dice).

Flank Attacks. A unit can only be hit in the flank when that unit is engaged frontally also. Attackers score casualties in 1st round of mêlée without defenders being able to fight back. If defender's morale allows, they fight back on 2nd round of mêlée.

A unit attacked in flank must be on the end of a line or in such a position as to have the normal width of a unit in free-space on that flank.

If unsuccessful—(a) Infantry when breaking carry on as in paragraph 3 of '*Cavalry attacking Infantry*'.

(b) Cavalry when breaking throw 1 dice per defender who then moves rearward $2\frac{1}{2}''$ for every 1 on the dice. Attackers automatically kill all the enemy they can reach within the limit of their move-distance.

Cavalry Fighting Cavalry

1. The defending unit throw dice, needing 4, 5 or 6 to neutralise attackers impetus. If successful, then they are considered to have counter-attacked and moved forward 3″ to meet their attackers. Mêlée then takes place in usual way.
2. If defending unit does NOT counter-attack—1 dice is thrown for each front-rank attacker:

Cavalry—5 or 6 takes out 1 defender—Saving throws of 4, 5 or 6.

The attacker then moves forward into the gap and takes on second rank man. If there is no second rank, then cavalry attacker goes through, deducts 3″ from his distance, turns and attacks from rear.

Infantry Fighting Cavalry

1. As in (1) above.
2. As in (2) above but infantrymen need 6 on each dice to take out a cavalryman.

Rear Attacks. Can only be launched from a point parallel to or in rear of the unit being attacked. This is excepted in the case of a cavalry attack on a unit that is pinned frontally—then the cavalry may ride around the unit and attack rearwards.

Method: As for Flank Attack, but defenders cannot fight back for *2* rounds of mêlée.

Pulling out of a Mêlée: After the two rounds of fighting in a single game-move, a force may pull out of that mêlée, withdrawing half their full move-distance in any reasonable direction providing it does not involve them in another mêlée.

Indians

Both sides employed Indians at various times, and the French usually had more Indian allies than the British. Those on the British side were the Five Nations (Mohawk, Seneca, Cayuga, Onondaga and Oneida) and the Stockbridge Indians.

Indians will be led by chiefs or white officers; one or more to each band. If a chief or leader is killed—or if the band suffers one third or more casualties, each man in group will have a dice thrown for him—score 1, 2 or 3 and he will abandon battle and leave the field. Indians may set-up ambushes in the following manner:

If a body of troops is moving down a road or path there exists the possibility of its being ambushed. To achieve this, the ambusher must in the first instance be in position at least one move-distance ahead of the troops.

If during this or subsequent moves scouts get within sight—3″—of the ambush then they will discover it. They may of course be attacked by the ambushers in hand-to-hand combat, but if they succeed in returning to the main column it is warned of the ambush.

If a scout sees the ambush but is attacked and killed, a dice is thrown to see if he fired a shot as a warning, 5 or 6 he has done. If the main column is within 12″ it will hear the shot and add 1 to its throw when entering the ambush area. (Wind direction can be taken into account if wished.)

If no warning reaches the main column, then it *MUST* continue down the road into the ambush. On arriving there, a dice is thrown, 1, 2, 3 or 4 the ambush works, 5 or 6 it is sprung too early and the column reacts in time.

If the ambush does not work then fire is exchanged in the normal way.

If the ambush works, the ambushers will fire a volley into the troops adding 50% to their score, who do not reply.

On the second move, the ambushers will fire a second volley into the troops, again adding 50%. The troops then dice for morale, deducting 2 from their throw. Those whose morale is bad will fire in the air, those with good morale may fire into the woods but get only 50% score.

On the following move those units which had bad morale will break and run. The remainder may advance, remain or retreat as wished. The ambushers may emerge and attack, follow up fugitives, etc., as they wish. Any of the runaways overtaken by an Indian is automatically killed.

MODEL SOLDIERS FOR WARGAMES

One of the features which makes it so attractive and convenient to wargame the American Revolution is the ready availability of model soldiers for that period. At the time of writing there is a considerable choice and it seems likely that the bi-centenary of the conflict will stimulate manufacturers to produce an even wider range.

A word on the subject of scales might aid the novice—the most common size of wargames figures are 25 mm which means that the little soldier is one inch in height. 00/H0 scale and 20 mm, more or less the same, are correspondingly smaller, while a 30 mm figure is about one and three-sixteenths of an inch, seemingly a small amount but producing an appreciably larger figure.

By far the cheapest and most readily available model soldiers in this period are the 00/H0 scale model soldiers made by Airfix of Great Britain. Their S39/59 box American War of Independence—Washington's Army (pictured in Fig. 4) is a 40-piece set which includes officers and men in a great variety of positions and even includes a mounted man who can be painted up as General George Washington himself. Moulded in authentic uniforms in a cream coloured plastic that readily lends itself to painting, the set consists of the following:

1 mounted officer: 2 dismounted officers: 1 Standard bearer with standard; 1 drummer; 1 man carrying gunpowder barrel; 6 men standing firing: 4 men crawling: 4 men lying firing: 4 men reloading: 6 men advancing: 6 men kneeling firing; group of 2 men carrying a wounded comrade.

As opponents, Airfix have produced the British Grenadier box S40/69 (shown in Fig. 5)—another 40-piece set of British Grenadiers in their distinctive caps.

The set consists of:

2 mounted officers; a Standard Bearer with standard; 1 drummer; 6 men kneeling firing; 6 kneeling reloading; 6 standing reloading;

6 men advancing; 6 men marching.

The figures from the Airfix 'Washington's Army' set need little converting and can be painted up to resemble the British or German infantryman of the 1776–81 period while the 'British Grenadiers' set, without alteration, can become the Grenadier Regiments of the various German States whose mercenaries were employed by Britain. By cutting heads with tricorne hats from the infantry figures and substituting them, cavalry can be converted from existing Airfix American Civil War cavalry or even their Napoleonic cavalry. It is even possible to cut a figure off at the waist and replace the top Napoleonic or American Civil War half with a suitable War of Independence section!

Every style of soldier in any position can be converted from one or other of the existing Airfix production figures. Requiring only a few basic tools, conversion of Airfix figures is an easy and cheap way to build up large armies of any desired type. Instructions on converting can be found in 'Military Modelling' and 'Better Military Modelling' by D. F. Featherstone (Kaye and Ward, London) and 'How to go Advanced Plastic Modelling' by Chris Ellis (Patrick Stephens, London, 1970). Illustrated articles on converting appear regularly in such British publications as *Airfix Magazine*; *Military Modelling*; *Slingshot* and *Wargamer's Newsletter*.

There are numerous reputable commercial manufacturers of metal soldiers with extensive ranges of figures suitable for the American Revolutionary period. Among the finest figures from a quality point of view is the very comprehensive 30 mm range made by the famous Charles Stadden and retailed by the British firm of Tradition, 5A and 5B Shepherds Street, London W.1. These figures are in many of the illustrations in this book. The American agent for Stadden and other Tradition figures is Polks Hobby Department Store, 314 5th Avenue, 32nd Street, N.Y. 1, New York, U.S.A., while A. C. Charles and Son, 3213 West Liberty Avenue, Pittsburg, Pa. 15216, U.S.A. stock the Tradition 25 mm range of wargames figures.

One of the best known commercial manufacturers of wargames figures in Great Britain is Miniature Figurines (28/32 Northam Road, Southampton SO2 0PA, Hampshire, England) who manufacture figures in perhaps the greatest variety of scales—5 mm (Regimental Blocks); 15 mm; 25 mm and 30 mm. American Agents:

S. Carpenter and G. Scott, Box P, Pine Plains, N.Y. 12567.

Miniature Figurines U.S.A. Inc., 4311 Lemmon Avenue, Dallas, Texas 75219.

Another well-known British manufacturer is Hinchliffe Models Ltd. (Meltham, Huddersfield, England HD7 3NX). Their American agents are Miniature Figurines U.S.A. Inc., 4311 Lemmon Avenue,

Dallas, Texas 75219, and Coulter Bennett Limited, 12158 Hamlyn Street, Suite 1, North Hollywood, California 916060.

An inexpensive range of 30 mm plastic figures are those made by Spencer-Smith Miniatures of 66 Long Meadow, Frimley, Camberley, Surrey, England. He makes infantry, cavalry and artillery for both American and British Revolutionary War forces.

Other British manufacturers who are known to have ranges of figures covering the American Revolution period are:

P. Laing, (15 mm scale only), 11 Bounds Oak Way, Southborough, Tunbridge Wells, Kent.

Willie Figures, (30 mm scale), 60 Lower Sloane Street, London SW3.

An American source of supply of wargames figures for the 1776–1781 period is Jack Scruby, the Soldier Factory, P.O. Box 1658, Cambria, California 93428, U.S.A. One of the first commercial manufacturers of wargames figures, Jack Scruby offers figures suitable for the American Revolution in three distinct scales—N gauge (9 mm); 25 mm and finally 30 mm scale. Then there is Custom Cast Inc., Independence Square, 57 South Main Street, Dayton, Ohio 45459, U.S.A.

All model soldiers are sold in the same pristine silver state as they emerge from the mould and they have to be painted. This necessitates having some knowledge of the different uniforms and their colourings. The wargamer will find his needs very amply supplied by the following volumes:

King George's Army 1775–1883—a handbook of British, American and German Regiments, by Philip R. N. Katcher (published in 1973 by the Stackpole Company, Cameron and Kalker Streets, Harrisburg, Pennsylvania 17105, U.S.A.).

American Soldiers of the Revolution, by Alan Kemp and by the same author *The British Army in the American Revolution* (both Almark Publications Great Britain).

George Washington's Army, by Brigadier Peter Young.

The American Provincial Corps 1775–1784, by Philip Katcher. (Both Osprey Men-at-Arms Series, Great Britain).

Uniforms of the American, British, French and German Armies in the War of the American Revolution 1775–83, by Lt. Charles M. Lefferts (New York 1926, reprinted in facsimile by WE Inc., 1970).

HOW TO PAINT YOUR MINIATURE ARMIES

AN ARMY OF well-painted figures marching across the table is a joy to behold and painting your soldiers is one of the most interesting and rewarding aspects of collecting and wargaming—but it is also the most difficult. To be able to paint well is a gift not granted to all—follow these simple instructions and anyone can be a reasonable painter.

The models should be prepared before being painted and any 'flash' or surplus metal attached to the figure must be filed or clipped away. Before putting on the top colours apply a base-coat or undercoat to the model, using an aerosol matt-white spray.

Good brushes are essential, size 00 and 000 are best for small work and short hair brushes are preferable to the longer types as they do not bend.

Model soldiers can be painted to a satisfactory standard with water colour, oil or plastic paints, depending upon whether the wargamer wishes his figures to have a flat (matt) or a gloss finish. Matt figures look more realistic but are not as durable and will 'gloss' or 'glaze' with the handling received whilst wargaming so the figures must be given a protective coating of clear flat lacquer or varnish. Matt water-colour paints are available in a good basic colour selection at reasonable prices; thinned with water, the colours dry rapidly and retain their brightness without fading. Painting with oil colours is a slow process because each colour must be allowed to dry before the next is applied. Some oil colours are transparent with less covering power and should be applied over an undercoat (usually white) otherwise the natural silver finish of the model shows through. Plastic-type paints in a vast range of colours and shades have reached a high standard, and are best for plastic model soldiers to prevent the paint flaking-off.

There are specific paints produced for military models in ranges which includes national military colours and shades such as gun-metal, brass, bronze, leather and camouflage paints.

Small 25 mm figures are more easily painted if one uses a magnifying glass mounted on a small stand so that the figure can be surveyed while both hands are left free. To aid painting, the little wargames figures should be glued, about 2 cms apart, to a thin strip of wood with enough of the wood left at one end to serve as a handle. When painting the wood is held in one hand or fixed in a spring clip on the edge of a table and both hands are free to paint.

The task of painting armies of figures has put many potential wargamers off the hobby so they must find a 'quick' method of painting figures to a satisfactory standard. Choose one of the main colours of the uniform, say the red of the tunic, and cover the top half of the figure with that colour—do not worry about covering the helmet, face or knapsack because each new colour is carefully painted along the line where it joins another colour. For example, when slapping on the red paint you are covering up the lapels and crossbelts which are white and the soldier's white trousers are marked with red paint. Never mind, when the red is dry, carefully paint in the white parts, taking great care with the line where the white and the red join.

Once the two main colours (red and white in this case) have been applied, paint the lesser colours—the grey blanket roll; the black helmet, shoes and pack; the blue cuffs and collar. The base on which the figure stands is the last part to be painted—bright green sets off the other colours; the face and hands next to last after the musket or weapon, which can be held whilst painting. The final finish of the figure is improved if the crossbelts, epaulettes, etc., are very carefully outlined in Indian ink with a fine-point pen, or with a very fine felt-tip pen.

Horses can be quickly painted by spraying them with a white undercoat and then paint them with a thinned-down black or brown. When dry, the white shows through the top colour to give the 'highlights' or sheen of a horse's coat. Basically, horses are—white (with brown to yellow/white mane and tail); brown (chestnut to dark brown—with dark brown mane and tail); light brown (very light brown to red brown—with mane and tail of brown with a touch of black) and black (with mane and tail of black with a touch of blue).

Some wargamers spend hours endeavouring to paint every single figure so that it is a collectors item although all that is required is a pleasing appearance in mass. In that way wargames figures, although not individually painted to a very high standard, present a pleasing appearance when formed up in armies.

BUILDING TABLE-TOP BATTLEFIELDS

WARGAMES ARMIES can form up on a bare table and march towards each other until they come into contact and begin the battle, but this is rather like staging a play against a single colour backcloth. Battles are not fought on open 'parade-grounds' lacking such tactical features as hills, woods, rivers and roads, although a good terrain is not necessarily a cluttered one and some of the best wargames are fought on a battlefield with a single hill or a clump of trees. The scenic features on a battlefield provide objectives and battles are won by capturing a hill, a village or a vital crossroads so that not only does an attractive terrain stimulate the wargamer but it also tests his tactical knowledge.

It is far more enjoyable to fight a battle on a natural-looking terrain than upon a table-top with books for hills, matchboxes for houses and chalked-in roads and rivers. Background scenery extended along the base lines and sides of the tables on 30 cms deep card or hardboard lengths will add the finishing touches to a good terrain.

The most realistic wargames terrain is a sand table—a large and strong table with sides about 20 cms high forming a large tray filled with fine plasterers sand to a depth of about 15 cms. The small model soldiers used for wargaming stand up quite well in sand, and as the sand is moistened when constructing the terrain, the table-top is lined with plastic sheeting. After dampening, the sand can be moulded with a plasterers trowel into hills, valleys, sunken roads, river-beds, trenches and shell-holes. With a large plasterers brush, lightly brush a thin mixture of coloured powder and water over the sand to represent green grass or brown plough and yellow cornfields. The roads are hollowed-out paths in the sand, giving the appearance of roadside verges, in the same way riverbeds are made with a watery mixture of green and blue poured into them, which runs down and settles to give the appearance of water. When trees, walls, fences, houses, bridges are placed in position, a very realistic countryside appears.

Undulations formed by a large green cloth stretched over books or pieces of wood makes a pleasing wargames terrain; coloured plastic sheet can be used and rivers and roads, ploughed fields, etc., painted in poster paint and washed off after the battle is finished. Slabs of wood or polystyrene piled upon each other make stepped hills, providing convenient contours on which soldiers can stand.

Hobby shops sell realistic trees or they can be made by sticking lichen moss onto pieces of twig. Stone walls and bridges can be purchased or made from balsa-wood or card. Houses of the Revolutionary period can be made up from commercial plastic kits or they can be scratch-built from plastic card, cardboard or balsa-wood.

The wargamer with a permanent room for his hobby can set up his battlefield at leisure and leave it so that the battle can continue for more than one session; but many wargamers have to erect and dismantle their wargames table on the same evening. They can assemble a very realistic terrain by using squares of hardboard or insulation board that fit in with each other; these scenic squares can be endlessly permutated to build up different battle terrains. On one square is a hill; a farm or village nestles on another whilst a third may have a river with a bridge; roads and rivers end at the same point on each square so that, when squares are put together, they can run across the table.

Inexpensive polystyrene ceiling tiles are easily cut and carved into roads, riverbeds, caves and shell-holes to form excellent scenic squares; and hills are made by having a large irregularly shaped piece for the base with shaped smaller pieces glued upon it. Polystyrene will take poster or water colour but so absorbs them that plenty of paint has to be used.

BIBLIOGRAPHY

THERE ARE A NUMBER of books that explain the hobby of wargaming, the best known being:

Little Wars, H. G. Wells (1913).
Wargames, Donald Featherstone (1962).
Wargames in Miniature, J. Morschauser (1963).
Naval Wargames, Donald Featherstone (1966).
Air Wargames, Donald Featherstone (1967).
Charge!, Brigadier P. Young and Lt-Colonel J. P. Lawford (1967).
Advanced Wargames, Donald Featherstone (1969).
Discovering Wargaming, John Tunstill (1969).
Introduction to Battle Gaming, Terence Wise (1969).
Battle! Practical Wargaming (*World War II*), Charles Grant (1970).
Battles with Model Soldiers, Donald Featherstone (1970).
Wargames Campaigns, Donald Featherstone (1970).
The Wargame, Charles Grant (1971).
Solo Wargames, Donald Featherstone (1972).
The Wargame, ed. Peter Young (1972).
Wargames Through the Ages, Vol. I—*Ancient and Mediaeval Periods*; Vol. II—1420–1783; Vol. III—1791–1859; Vol. IV—1861–1945, Donald Featherstone (1972–76).
Battle Notes for Wargamers, Donald Featherstone (1973).
Tank Battles in Miniature—The Western Desert Campaign, Donald Featherstone (1973).
Napoleonic Wargaming, Charles Grant (1973).
Practical Wargaming, C. F. Wesencraft (1974).
Napoleonic Wargaming, Bruce Quarrie (1974).
The Napoleonic Wargame, G. W. Jeffery (1974).
With Pike and Musket, C. F. Wesencraft (1975).
Ancient Wargaming, Phil Barker (1975).
Tank Battles in Miniature—The Russian Campaign 1941–45, Bruce Quarrie (1975).
Skirmish Wargaming, Donald Featherstone (1975).

Wargaming Ancient and Medieval Periods, Donald Featherstone (1976).

Wargaming Pike and Shot Period, Donald Featherstone (1976).

Both in America and Great Britain there are magazines and journals concerned with the hobby of wargaming. The longest established and best known is *Wargamer's Newsletter* (Tradition, 5A and 5B Shepherd Street, London W.1, England).

Then there are:

Arquebusier, Journal of the Pike and Shot Society—Duncan Macfarlane, 23 Nayfield Close, Driffield, North Humberside, England.

The Bulletin, The British Model Soldier Society, 16 Charlton Road, Kenton, Middlesex, England.

The Courier, Bulletin of the New England Wargamers Association —R. Bryant, 45 Willow St., Brockton, Mass. 02401, U.S.A.

Savage and Soldier, Ian N. Cohen, 13426 Kingscross Lane, St. Louis, Mo., U.S.A. 63141.

Slingshot, Journal of the Society of Ancients—John L. Norris, Hillside, Jackson's Lane, Highgate Village, London N6 5SR.

Tradition, 5A and 5B Shepherd Street, London W.1, England.

Books on the War of the American Revolution

Alden, J. R., *General Gage in America* (1948).

Anderson, T. S., *The Command of the Howe Brothers during the American Revolution* (1936).

Bass, R., *The Green Dragoon* (1957).

Bill, A. H., *Valley Forge, The Making of an Army* (1952).

Blien, B., *Battle for Manhattan* (1955).

Boatner, M. M., *Encyclopedia of the American Revolution* (1966).

Bowen, C. D., *John Adams and the American Revolution* (1950).

Bridenbaugh, C., *Cities in Revolt* (1955).

Brown, W., *The Good Americans* (1969).

Brown, W., *The King's Friends* (1965).

Callahan, N., *Flight From the Republic* (1967).

Carman, W. Y., *British Military Uniforms* (1957).

Clinton, Sir Henry, *The American Rebellion: Sir Henry Clinton's Narrative* (1954).

Commager, H. S. and Morris, R. B., *The Spirit of '76* (1958).

Cullen, J. P., *The Concise Illustrated History of the American Revolution* (1972).

Cuneo, J. R., *Robert Rogers of the Rangers* (1959).

Curtis, E. E., *The Organisation of the British Army in the American Revolution* (1972).

Darling, A., *Red Coat and Brown Bess* (1968).

Davis, G., *Regimental Colours in the American Revolution.*

Dorson, R. M., *American Rebels: Narratives of the Patriots* (1953).

Dupuy, R. E., *The Compact History of the U.S. Army* (1961).

Dupuy, R. E., *Men of West Point* (1951).

Dupuy, R. E., *The Battle of Hubbardtown* (1960).

Dupuy, T. N., *The Encyclopaedia of Military History.*

Dupuy, T. N. and Hammerman, G. M., *The Military History of Revolutionary War Land Battles* (U.S.A. 1970).

Eggenberger, D., *A Dictionary of Battles* (1967).

Esposito, V. J., *The West Point Atlas of American Wars* (1959).

Flemming, T., *Now We are Enemies* (1960).

Fortescue, John W., *A History of the British Army* (1930).

Freeman, D. S., *George Washington* (1954).

French, A., *The Day of Concord and Lexington* (1925).

French, A., *General Gage's Informers* (1932).

French, A., *The First Year of the American Revolution* (1934).

Fuller, Maj-General J. F. C., *Decisive Battles of the United States of America* (1942).

Galvin, J., *The Minute Men* (1967).

Ganoe, W. A., *The History of the United States Army* (1942).

Graves, C., *Sergeant Lamb of the 9th* (1940).

Graves, C., *Proceed Sergeant Lamb* (1941).

Harbottle, T. B., *A Dictionary of Battles* (Reprinted 1966).

Heinl, R. D., *Soldiers of the Sea—The United States Marine Corps 1775–1962* (1962).

Johnson, Curt, *Battles of the American Revolution* (1975).

Katcher, P. R. N., *King George's Army 1775–1783* (1973).

Kemp, Alan, *American Soldiers of the Revolution* (1972).

Ketcham, R. M., *The American Heritage Book of the Revolution* (1958).

Klinger, R. L. and Wilder, R. A., *Sketchbook '76, The American Soldier 1776–83*

Lancaster, B., *From Lexington to Liberty* (1955).

Lancaster, B., *The American Heritage Book of the Revolution.*

Lawson, C. C. P., *A History of the Uniforms of the British Army* (1960).

Lefferts, C., *A History of the Uniforms of the American, British, French and Germans in the War of the American Revolution, 1775–1783* (1926).

Lossing, B. J., *The Pictorial Field Book of the Revolution 1775–1783* (1885 reprinted).

Macksey, P., *The War for America* (1965).
McMunn, G., *The American War of Independence in Perspective* (1939).
Miller, J. C., *Origins of the American Revolution* (1943).
Miller, J. C., *Triumph of Freedom 1775–1783* (1948).
Millis, W., *Arms and Men* (1956).
Mitchell, J. B., *Decisive Battles of the American Revolution* (1962).
Montross, L., *Tag and Bobtail* (1952).
Montross, L., *The Reluctant Rebels* (1950).
Montross, L., *War Through the Ages* (1944).
Morison, S. E., *The Oxford History of the American People* (1953).
Morris, R. B., *Encyclopedia of American History* (1953).
Morris, R. B., *The American Revolution Reconsidered* (1967).
Ogden, H. A., *Army of the U.S.A.*
Peckham, H. H., *The War for Independence* (1958).
Roberts, K., *The Battle of Cowpens* (1958).
Scheer, G. F. and Rankin, H. F., *Rebels and Redcoats* (1957).
Scott, J. A., *The Diary of the American Revolution* (1967).
Sears, R., *Pictorial History of the American Revolution.*
Smith, P., *Loyalists and Redcoats* (1964).
Spaulding, O. L., *The United States Army in War and Peace* (1937).
Sweetman, J., *Saratoga 1777.*
Swiggett, H., *The Forgotten Leaders of the Revolution* (1955).
Todd, F. P. and Kredel, F., *Soldiers of the American Army* (1954).
Tucker, G., *Poltroons and Patriots* (1954).
Wade, H. and Lively, R., *This Glorious Cause* (1958).
Wallace, W. M., *Appeal to Arms* (1951).
Ward, C., *The War of the Revolution* (1952).
Wickwire, F. and M., *Cornwallis* (1967).
Wickwire, F. and M., *Cornwallis and the War of Independence* (1971).
Young, P., *George Washington's Army* (1972).
Young, P., (ed.) *The Wargame* (1972).

INDEX